THE GAS TURBINE

THE GAS TURBINE

BY NORMAN DAVEY

NEW YORK
D. VAN NOSTRAND CO
TWENTY-FIVE PARK PLACE
1914

PREFACE

A PREFACE may be considered, in the first place, to be something in the nature of an apology on behalf of the author for helping to swell the great number of books with which the reading public are so unduly pestered. Unless the author feels sincerely that he has something to say worth the saying he had better avoid the printing press as he would the devil. For, in such a case, he will be doing a disservice to two classes of people. First, he will narrow the market for other—and, perhaps, more worthy—bookmakers :—

> " Thou shalt not covet, but tradition
> Approves all forms of competition."

And, secondly, he will be burdensome to such readers—and, of these there are not a few—who believe that—

> " A book's a book although there's nothing in't."

As to whether there is anything in the volume to which these pages form a preface, it is not my privilege to say. That is for the reader to determine. What I may be permitted to do is to excuse myself of the first of these two disservices, namely, to plead that I have not dumped my goods upon a glutted market.

As far as I am aware, there are only two books extant in the English tongue on the Gas Turbine. The first of these consists of a number of scientific papers, bound together, and edited by an American in 1909. Though a useful reference book, containing, moreover, the account of the Armengaud-Lemale experiments taken from *Cassier's Magazine,* it is now out of date, inasmuch as the excellent work of Herr Holzwarth was effected since its publication. The other book is the English translation of Herr Holzwarth's account of his gas turbine at Mannheim in 1911. While the insufficiency of Mr. Suplee's book is that it has nothing to say about the Holzwarth turbine, the inadequacy of Herr Holzwarth's book is that it has nothing to say about anything else ; it is, in fact, a monograph (and a very valuable monograph) dealing with the operation of one particular machine. Of books on the gas turbine in other languages than English there are (besides the German edition of Holzwarth), in French, M. Ventou-Duclaux's "Les Turbines à Gaz" (Dunod et Pinat, Paris, 1912), and, in Italian, Ing. Egidio Garuffa's "Le Turbine a Gas" (Bietti and Reggiani, Milan, 1913), neither of which volumes gives a comprehensive survey of the gas turbine problem. It therefore seemed to me that a book dealing with the gas turbine in a comprehensive and a catholic manner, written in the English tongue, would not be in the nature of an attempt to fill a want already supplied.

The subject falls naturally into the two divisions of theory and practice ; at the present stage of research it is inevitable that the former should preponderate over the latter, though it cannot be too often emphasised that the gas turbine problem is wholly a practical one. Of the three types of gas turbines —the constant-pressure, single-fluid ; the constant-volume, single-fluid ; and the constant-pressure, mixed-fluid, the last named has not up to now received the notice it deserves. I have here attempted to remedy that neglect.

I have adopted the more commonly accepted values for thermodynamic constants in all calculations ; a detailed analysis of the variations of these with external conditions is given in Chapter VI. To burden all formulæ with the interpolation of variable constants in order to eliminate an "error" of one in three hundred is simply rabbinical.

The Centigrade scale of temperature has been adopted throughout this book, the thermal unit used being that quantity of heat required to raise one pound of water through 1° C.—after the precedent of Sir James Ewing.

The second purpose of a preface is to give an opportunity to the author of thanking such as have aided him in his undertaking. Let me then acknowledge my gratitude to *The Engineer* for permission to make use of matter that I contributed to that periodical in the spring of 1912 ; to Herr Hans Holzwarth, for information, kindly and courteously given, and to Messrs. Oldenbourg, for the loan of their blocks ; to Mr. P. J. Mitchell, for details concerning the Rateau pump ; to *Cassier's Magazine*, for permission to reproduce Fig. 92 ; and to Messrs. Escher, Wyss & Co., for illustrations of the Zoelly compressor.

I have added in an Appendix a list of the Gas Turbine Patents in the English Patent Office from the year 1856 to September, 1913.

NORMAN DAVEY.

EWELL, SURREY.
January, 1914.

CONTENTS

LIST OF ILLUSTRATIONS

THE GAS TURBINE

BOOK I. THE THEORY

CHAPTER I. GENERAL CONSIDERATIONS OF THE GAS TURBINE AS A HEAT ENGINE

THE gas turbine, in the broadest sense in which that term can be used, may be divided into three classes of rotary machines. All these three types use the expansion of a gas for the production of work translated into rotative motion without the use of intervening reciprocating parts. This is the largest sense in which the word "turbine" can be applied; it is not the absolute, accurate, and limited sense in which the word should be used. Of late, however, a method has been advocated of employing the expansive power of gases *indirectly* to the production of rotary motion, which cannot well be neglected in any book dealing with the gas turbine. It is therefore necessary to regard the gas turbine, first of all, in its broadest basis of definition.

The three methods whereby the expansive power of a gas may be utilised to produce rotary motion without the interpolation of reciprocating parts are as follows :—

First, the so-called "fixed abutment" type of rotary explosion motor. This type of machine has never been a success, though numberless patents have been taken out for engines of this kind, more especially during the last two decades of the nineteenth century. The system is now practically defunct and need not be considered any further here. The machine is of the rotary piston order, and does not come into the stricter sense of the word "turbine" at all.

Secondly, the gas-driven water turbine. This special type of machine has come to the front of late in connection with the gas explosion pump of Mr. Humphrey. The expansion of gas is made to give kinetic energy to a mass of water, which energy is absorbed by a turbine wheel. It is interesting to note that one of the earliest patents for a gas turbine was for a machine working on this principle, viz., Newton, No. 1672, in the year 1865. This type of rotary machine, though a *turbine* in the accurate sense, can hardly be considered as a *gas turbine* in the accurate sense. It will be considered, briefly, in a subsequent chapter.

Thirdly, the gas turbine, properly so named. This machine may be absolutely defined as " a machine in which the potential energy of a gas, or a mixture of gases, or a mixture of gas and vapour (but not a vapour alone), is converted into kinetic energy ; and the energy thus produced, absorbed by rotary mechanism without the intervention of reciprocating parts." This is the gas turbine as generally understood since Barber's patent in 1791 to the present day.

The gas turbine consists essentially of three integral parts : the furnace, or heat source ; the turbine (or turbine and pump), or energy-absorber ; and the cooler (or condenser) or heat-sump. Gas turbines may be divided into three classes, according to the cycle on which they operate. The cycles in question can be correlated with their congeners in internal combustion piston engines. The three cycles are as follows :—

CYCLE I.—Heat taken in at constant pressure.

In gas engines : The Diesel cycle.
In gas turbines : The Armengaud and Lemale cycle.

CYCLE II.—Heat taken in at constant volume.

In gas engines : The Otto (or Lenoir) cycle.
In gas turbines : The Holzwarth cycle.

CYCLE III.—Heat taken in at neither constant pressure nor constant volume, but during thermodynamic conditions varying between these states.

In gas engines : There is no theoretical cycle of such a nature among piston engines ; in practice, however, the Otto cycle tends to take in heat at a slightly varying volume (*i.e.*, the " explosion line " on the diagram leans forward slightly ; with the old Lenoir engine this was much more marked).

In gas turbines : The Karavodine cycle.

Of these three cycles (onall of which actual gas turbines have been constructed and put into operation) only the first two have given anything like rn efficiency approaching that obtained with other heat engines. The third cycle is essentially inefficient; it invites, however, extreme simplicity of construction.

THEORETICAL ANALYSIS OF THE THREE CYCLES.

Cycle I.—Heat Taken in at Constant Pressure.

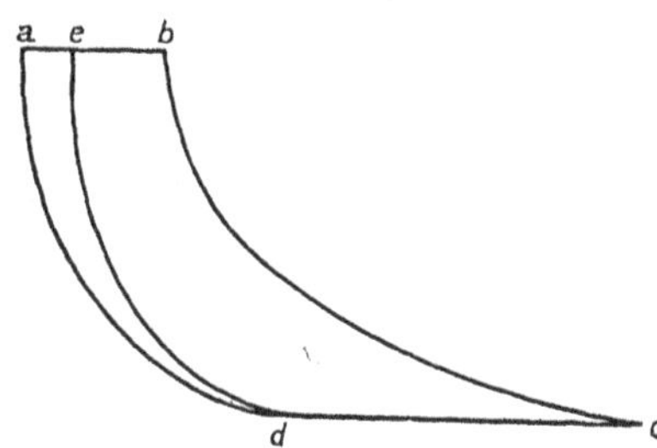

FIG. 1.—"PV" Diagram for Cycle I.

In the "PV" diagram (Fig. 1) the gas is heated at constant pressure from *a* to *b*; it is then expanded adiabatically in the turbine from *b* to *c*, the temperature dropping in the process; heat is rejected to the cooler from *c* to *d*, at constant pressure, and the fluid is compressed to its original pressure, isothermally along *da*, or adiabatically along *de*.

The same series of operations is shown in the entropy-temperature diagram (Fig. 2). *ab* represents the absorption of the heat at constant pressure; *bc*, the adiabatic expansion; *cd*, the rejection of heat at constant pressure; *da*, the isothermal compression, or *de*, the adiabatic.

The efficiency of any cycle is :—

$$\frac{H - h}{H}$$

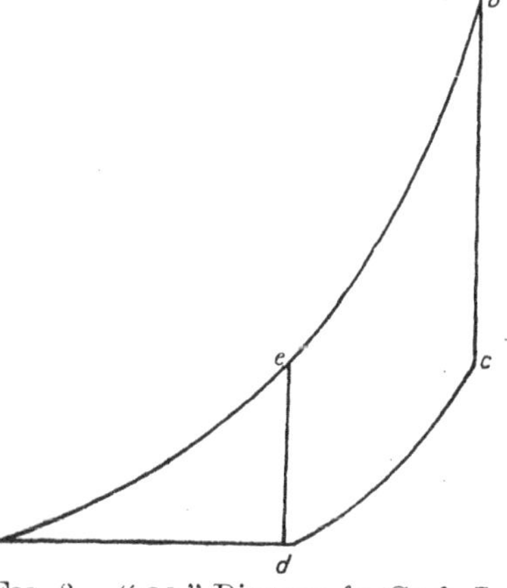

FIG. 2.—"ΘΦ" Diagram for Cycle I.

where H equals the heat put into the system, and *h* the heat rejected from the system. For Cycle I. the efficiencies are deduced as follows :—

(A) With isothermal compression (Armengaud and Lemale system) :

$$\eta = \frac{H - h - h'}{H}$$

where h' = negative work (expressed as heat units) done on pump

$$= \frac{C_p(\theta_1 - \theta_0) - C_p(\theta_2 - \theta_0) - R\theta_2 \log r_\epsilon}{C_p(\theta_1 - \theta_0)}$$

where θ_1 is the temperature at b; θ_2 the temperature at c; θ_0 the temperature at a and d; r is the compression "ratio" (pressure), and R is the constant for the gas equation.

Then $$\eta = \frac{(\theta_1 - \theta_2) - \theta_2 \log \frac{\theta_1}{\theta_2}}{\theta_1 - \theta_0}$$

substituting the value $C_p\left(1 - \frac{1}{\gamma}\right)$ for R.

(B) With adiabatic compression (Diesel system) :

$$\eta = \frac{H - h}{H} = \frac{C_p(\theta_1 - \theta_3) - C_p(\theta_2 - \theta_0)}{C_p(\theta_1 - \theta_3)}$$

$$= 1 - \frac{\theta_2 - \theta_0}{\theta_1 - \theta_3}$$

$$= 1 - \left(\frac{1}{r}\right)^{\gamma - 1}$$

where θ_1 is the temperature at b; θ_2 the temperature at c; θ_3 the temperature at e; θ_0 the temperature at d; and r the ratio of initial to final pressure.

In the Diesel oil engine, the heat is rejected at constant volume instead of at constant pressure, in which case the theoretical efficiency of the cycle becomes—

$$\eta = 1 - \frac{\theta_4 - \theta_0}{\gamma(\theta_1 - \theta_3)}$$

where θ_4 is the temperature at which the working fluid begins to reject heat to waste.

Cycle II.—Heat taken in at Constant Volume.

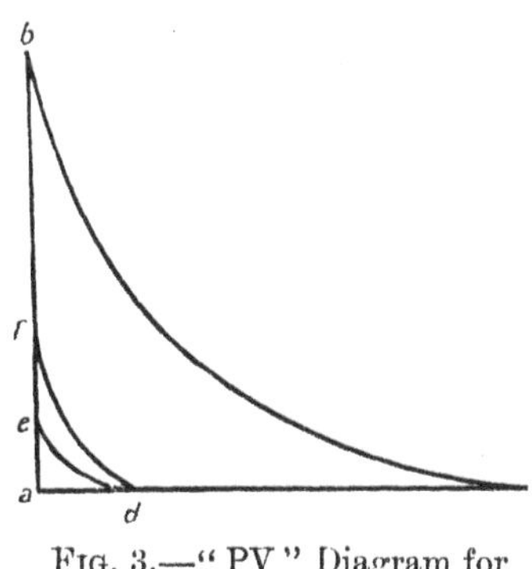

FIG. 3.—" PV " Diagram for Cycle II.

The gas is heated in a closed vessel, so that the volume is constant, by internal combustion (or otherwise); it is then allowed to expand adiabatically through a turbine, dropping in temperature during the process; heat is rejected at constant pressure to the cooler.

The working fluid may be initially compressed—adiabatically (as in the Otto cycle), or isothermally; or there need be no initial compression (as in the Lenoir cycle). Fig. 3 represents a " PV " diagram for the cycle.

The fluid is heated at constant pressure from *a* (or *e*, or *f*) to *b*; is expanded adiabatically from *b* to *c*; heat is rejected at constant pressure from *c* to *a* (or *d*); initial compression may take place from *d* to *e* (isothermal) or from *d* to *f* (adiabatic).

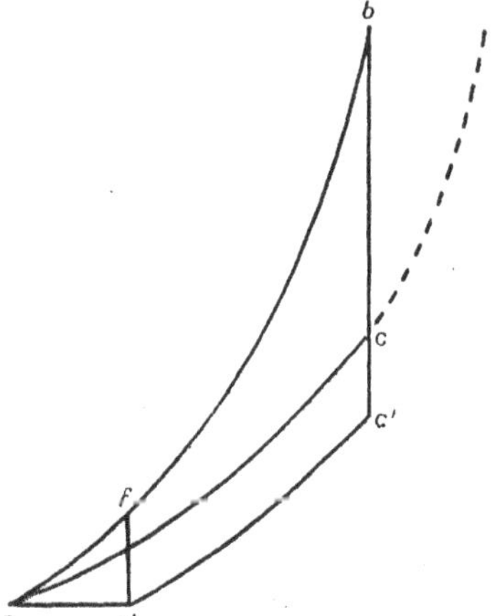

FIG. 4.—" ΘΦ " Diagram for Cycle II.

An entropy-temperature diagram for the same cycle is shown in Fig. 4. Heat is absorbed at constant volume along *ab*; adiabatic expansion occurs along *bc* (*c'*); heat is rejected at constant pressure along *ca*; or, in case of initial compression, *c'd*. *da* represents isothermal compression; *df*, adiabatic.

Efficiencies.—(A) No initial compression (Lenoir system):

$$\eta = \frac{H - h}{H} = \frac{C_v(\theta_1 - \theta_0) - C_p(\theta_2 - \theta_0)}{C_v(\theta_1 - \theta_0)}$$

$$= 1 - \frac{\gamma(\theta_2 - \theta_0)}{(\theta_1 - \theta_0)}$$

$$= 1 - \frac{\gamma(r^{\frac{1}{\gamma}} - 1)}{r - 1}$$

where θ_0 = temperature at a; θ_1, temperature at b; θ_2, temperature at c; and r, the ratio of the initial to the final pressure.

(B) With initial compression.

I. Isothermal compression (Holzwarth system):

Let temperature of the fluid at e, b, and c, be θ_0, θ_1, and θ_2, respectively: then—

$$\eta = \frac{H - h - h'}{H} = \frac{C_v(\theta_1 - \theta_0) - C_p(\theta_2 - \theta_0) - R\theta \log r_p}{C_p(\theta_1 - \theta_0)}$$

where r_p = compression "ratio" of pump (pressures).

Then $$\eta = \frac{\frac{1}{\gamma}\theta_1 - \theta_2 + \frac{\gamma - 1}{\gamma}\left\{\theta_0 - \log_e r_p\right\}}{\theta_1 - \theta_0}$$

II. Adiabatic compression (Otto system):

Let temperatures at d, b, c, and f, be θ_0, θ_1, θ_2, and θ_3, respectively: then—

$$\eta = \frac{H - h}{H} = \frac{C_v(\theta_1 - \theta_3) - C_p(\theta_2 - \theta_0)}{C_v(\theta_1 - \theta_3)}$$

$$= 1 - \frac{\gamma(\theta_2 - \theta_0)}{(\theta_1 - \theta_3)}$$

$$= 1 - \frac{r^{\frac{1}{\gamma}}/r_p^{\frac{\gamma-1}{\gamma}} - 1}{r - r_p^{\frac{\gamma-1}{\gamma}}}$$

Cycle III.—Heat taken in at neither Constant Pressure nor Constant Volume, but under Conditions varying between these States (Karavodine system).

This condition occurs with explosion in an open vessel, or in a vessel with only a partially closed exit. These conditions were approximated in the Karavodine turbine. The efficiency was essentially low.

Fig. 5 is a "PV" diagram for this cycle. The working fluid takes in heat along *ab*, while changing both its pressure and its volume. Adiabatic expansion takes place through the turbine along *bc*; rejection of heat at constant pressure along *cb*.

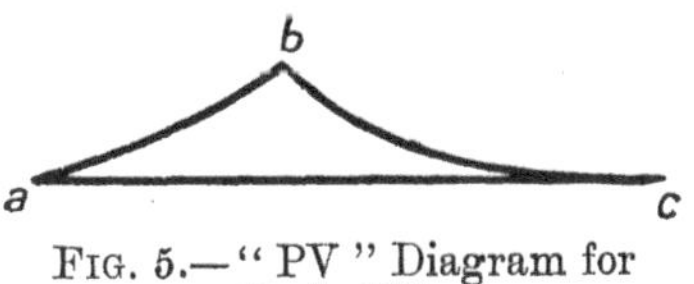

FIG. 5.—"PV" Diagram for Cycle III.

In the temperature-entropy diagram (Fig. 6) the absorption of heat takes place along *ab*, an arbitrary line, lying between the lines of heat-absorption at constant pressure (*ap*) and heat-absorption at constant volume (*av*). Adiabatic expansion takes place along *bc*, and rejection of heat at constant pressure along *ca*.

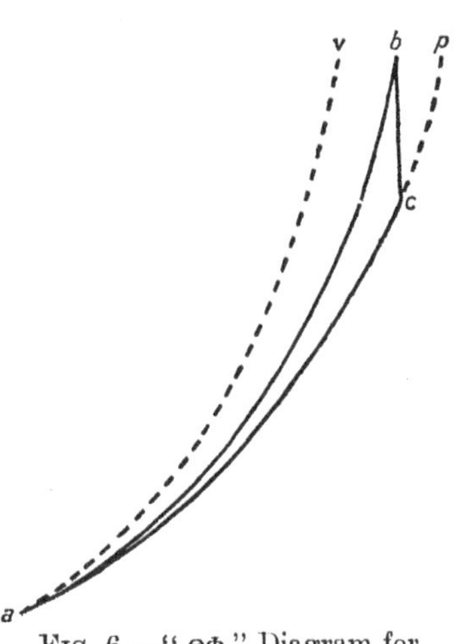

FIG. 6.—"ΘΦ" Diagram for Cycle III.

Efficiency.—Heat taken in at some state between constant pressure and constant volume (Karavodine).

Let temperatures at *a*, *b*, and *c*, be θ_0, θ_1, and θ_2, respectively. Then—

$$\eta = \frac{(C_p \text{ to } C_v)(\theta_1 - \theta_0) - C_p(\theta_2 - \theta_0)}{(C_p \text{ to } C_v)(\theta_1 - \theta_0)}$$

$$= 1 - \frac{\gamma(\theta_2 - \theta_0)}{(1 \text{ to } \gamma)(\theta_1 - \theta_0)}$$

It has already been mentioned that gas turbines may use

either a single gas or a mixture of gas and vapour ; the thermodynamic cycle is in either case the same. It is, however, convenient to separate gas turbines into two large classes, according as they use a gas only, or a gas and vapour together, as their working fluid. The former are known as *single-fluid* turbines ; the latter as *mixed-fluid* turbines. Single-fluid turbines are workable on the three cycles already described. Mixed-fluid gas turbines are theoretically possible in each of the said three cycles, but the dilution of the gas with a vapour (*i.e.*, steam) is only thermodynamically advantageous in the cycle in which the heat is absorbed at constant pressure ; that is to say, in turbines of the Armengaud-Lemale variety. From the purely theoretical valuation of the cycle, the efficiency is lowered by any addition of steam to the gaseous fluid, but in actual practice there is a considerable gain in economy by so doing. Limits of temperature, pressure, and peripheral speed, together with the inefficiencies inherent in pump and turbine, reduce the efficiency of the machine to a degree such that the addition of steam (under the conditions of superheat, inaugurated by its injection into the products of combustion) is of considerable economic value. It is also of great utility in reducing the temperature of the hot gases on the turbine wheel to a limit compatible with the material of which the blades are made. The practical consideration of this (and other) cycles will be discussed in detail later. The cycle of thermodynamic operations in a mixed-fluid turbine using air (products of combustion) and steam is shown in Figs. 7 and 8. The diagrams are not drawn to scale, and the proportion of the steam to the air diagrams is purely arbitrary. Fig. 7 is a " PV " diagram for the system : *abcd* is the diagram for the

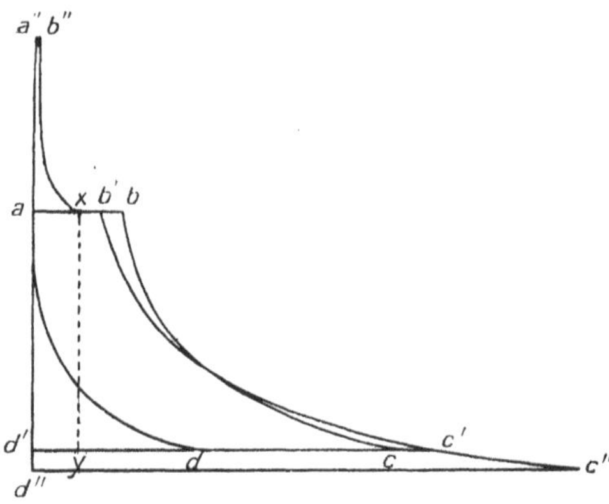

Fig. 7.—" PV " Diagram for Mixed-Fluid Turbine.

air, and $a''b''b'c'd'$ is the diagram for the steam superimposed upon that of the air. The steam may be raised to a pressure, higher than that of the air in its initial state, in a separate vessel, and then expanded through a separate turbine into the heat-source of the gas turbine proper, further expansion taking place to the final pressure in the heat-sump, condenser, or cooler. This initial expansion of the steam is represented in Fig. 7 by the curve $b''x$; the latter expansion, coincident with that of the air, by the curve $b'c'$; further expansion of the steam may, of course, be effected by condensation of the same in a separate condenser (the steam having been expanded in a separate wheel); such an operation is shown by the curve $c'c''$, and the subsequent condensation by the base-line, $c''d''$. The increase in volume of the steam from x to b' represents the effect of the superheat.

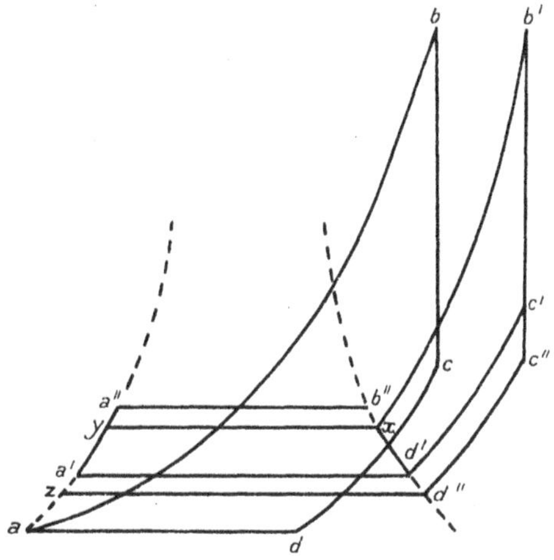

FIG. 8.—" ΘΦ " Diagram for Mixed-Fluid Turbine.

An entropy-temperature diagram for the steam-and-air system is shown in Fig. 8: $abcd$ is the diagram for the air; $a'a''b''b'c'$ (c'') d' (d'') that for the steam. The extra work done by the steam due to expansion above the upper limit of the air pressure is shown by the area $a''b''xy$; the work due to the superheat of the steam by the heated gases of combustion, the area $xb'c'd'$; the extra work done by expansion below the lower limit of pressure in the air turbine, the area $a'd'c'c''d''z$.

The addition of steam to the gas turbine is possible in any proportion, provided the temperature of the working fluid is not fixed; the more steam added, of course, the less the temperature. There neither is, nor can be, any line of demarcation between the case in which a minimum quantity of steam is used on the turbine wheel to counteract over-heating, and the mixed-fluid turbine in which the proportion of steam added approaches that of the air used, and the ultimate efficiency of the system is materially altered. It is, nevertheless, convenient to regard the "steam and air" by itself; the type in which small quantities of vapour are added to the working fluid being considered as a single-fluid turbine.

The gas turbine, indeed, can be regarded as being the product of three distinct lines of evolution from pre-existing types of heat engines. First, the constant-pressure, single-fluid gas turbine. This is simply the evolved product of the air engine in which a turbine wheel replaces the engine cylinder, and a pump (rotary or otherwise) the pump cylinder.

Secondly, the steam and air turbine. This approaches to a steam turbine in which the burnt products of combustion are expanded through the turbine together with the steam. The heat here is absorbed at constant pressure.

Thirdly, the explosion turbine. This is evolved from the gas engine of the Otto (or Lenoir) type, heat being taken in at constant volume and rejected at constant pressure. The Karavodine type of gas turbine, in which the heat is absorbed at a state varying between that of constant volume and constant pressure, is also here included.

The further consideration of the gas turbine will be treated under such classification, viz., the constant-pressure single-fluid gas turbine, or "gas" turbine; the constant-pressure mixed-fluid turbine, or "steam and gas turbine"; and the constant-volume gas turbine, or "explosion" turbine.

The cycles so far considered admit only of adiabiatic expansion in the turbine; it is possible, of course, to conceive a turbine unit working with isothermal expansion, by a judicious system of heating. No such scheme is capable of being carried out in practice owing to the difficulty of getting heat through metal walls to act upon so bad a heat-conductor as air. But such systems are interesting, as, with isothermal expansion, it is possible to construct a gas turbine working on the Carnot cycle. The efficiency would then, of course, be $\frac{T_1 - T_2}{T_1}$; such cycles will be discussed more fully at the end of the succeeding chapter.

CHAPTER II. THE CONSTANT-PRESSURE, SINGLE-FLUID GAS TURBINE

ALL the gas turbines of this class are composed of four essential units: the furnace, or heat source; the turbine, performing the positive work of the system; the pump, performing the negative work of the system; and the cooler, or heat-sump. The cooler may be used in a partially regenerative fashion, the supply of fluid to the furnace, abstracting a certain quantity of heat from the exhaust from the turbine. In two particular types of turbines of this class the heat-sump is simply the atmosphere. The absolute heat efficiency of such a cycle is, as already stated, given by the general formula—

$$\eta = \frac{H - h - h'}{H}$$

where H is the heat entering the cycle, h the heat rejected in the cooler, and h' the heat rejected in the pump if isothermal compression is used; with adiabatic compression h', of course, becomes zero.

Before considering the question of efficiency any further it may as well be made quite clear as to the exact manner in which that term is to be used, and how it is to be qualified. Perhaps more misunderstanding has come about over the misinterpretation of the word "efficiency," or the words "thermal efficiency," than upon any other matter of pure definition in any branch of science. A case, very much in point, was the general misinterpretation by the technical press of the (somewhat involved) statements made by Herr Holzwarth as to the efficiency obtained in his explosion turbine. In the ensuing pages the following are the ways in which the term "efficiency" will be used:—

THERMAL EFFICIENCY.

By "thermal efficiency" is meant the ratio of the quantity of heat available for work to the total quantity of heat put into the system ; this value only includes those losses that are inherent in the cycle used and consequent upon the temperatures between which the cycle operates.

This ratio is denoted by the symbol " η."

THERMODYNAMIC EFFICIENCY.

By " thermodynamic efficiency " is meant the ratio of the quantity of heat appearing as work done on the engine shaft to the quantity of heat available for work on the engine shaft. When applied to the *turbine* it represents the ratio of heat appearing as work done on the *turbine* shaft to the quantity of heat available for work on the turbine shaft. When applied to the *pump*, it represents the ratio of the quantity of heat appearing as work done by the pump in *isothermal* compression or exhaustion) to the quantity of heat expended in the form of work done *on* the pump by the engine or turbine driving it.

These ratios are denoted by the symbols " e," " e_t," " e^p," as applied to heat engine, turbine, or pump respectively.

OVERALL EFFICIENCY.

By " overall efficiency " is meant the ratio of the quantity of heat appearing as available, external work on the turbine shaft, or work done *on* the driven unit by the turbine driving unit, to the quantity of heat entering the furnace or heat source ; it being assumed that no thermal losses occur in the " furnace," or during the translation of heat energy to kinetic energy in the nozzle or nozzles provided for that purpose, and, also, that no heat exchange takes place between the working fluid and the walls of the vessels through which it passes (exclusive of the heat exchange in the cooler and regenerator), and that there are no radiation losses throughout the system. This value includes the loss coincident upon the difference in temperature between the inner and outer walls of the regenerator necessary for the flow of heat through them, and similar losses with externally-fired furnaces.

This ratio is denoted by the symbol " E."

FUEL ECONOMY.

By " fuel economy " is meant the ratio ot the work appearing on the turbine shaft, available for external use, in *brake horse-*

power per unit of time, or the work done on the driven unit by the driving unit in brake horse-power per unit of time, to the actual weight of coal (or coke, peat, wood, oil, etc.) supplied to the gas producer (or combustion chamber, furnace, burner, atomiser, etc.) per unit of time. This value includes *all* losses between the "furnace doors" and the turbine shaft. The ratio is expressed in lbs. of coal (or coke, peat, wood, oil, etc.) per brake horse-power per hour.

The ratio is denoted by the symbol "FE."

The word "efficiency," when used alone, is to be interpreted "thermal efficiency" ("η"). The use of the words "mechanical efficiency" has been thought to be possible of misconception; it has been avoided as far as possible. In cases where it has been used (*i.e.*, the "mechanical efficiency" of a turbine blade, etc.) the meaning is not open to misconstruction.

All constant-pressure gas turbines operate on one of two cycles, according as the working fluid is compressed isothermally or adiabatically upon its exit from the turbine. The fluid, at temperature T_0, pressure p_1, and volume v, takes in heat at constant pressure until the condition $T_1p_1v_1$ is reached. The gas is then expanded adiabatically through the turbine until the temperature drops to T_2; the pressure being p_2, and the volume v_2. The gas is then cooled in the cooler until the temperature reaches a value T_0, the pressure p_2, and the volume v_3. The gas is then compressed isothermally until the pressure becomes p_1, and the volume v.

If the compression is adiabatic the system takes in heat at constant pressure from a condition of the working fluid T_3p_1v, to a condition $T_1p_1v_1$. Adiabatic expansion occurs to a temperature T_2, the pressure being p_2 and the volume v_2. Heat is rejected to the cooler from $T_2p_2v_2$ to $T_0p_2v_3$. The fluid is then compressed adiabatically until the pressure equals p_1, the temperature being T_3, and the volume v.

In dealing with the question of the efficiency of the constant-pressure single-fluid gas turbine, the thermodynamic efficiency of the turbine and of the pump employed to compress the fluid to its original pressure is a matter of the greatest instancy to the economy of the plant. With the steam turbine, and all heat engines in which there is no "negative work," the overall efficiency of the whole machine is simply the multiple of the thermal efficiency of the cycle used and the thermodynamic efficiency of the turbine or engine. Or—

$$E = \eta \cdot e.$$

In heat engines, however, in which the negative work bears anything approaching a noticeable proportion of the total work, the formula—

$$E = \eta \cdot e$$

no longer holds.

Suppose, for example, that the thermodynamic efficiency of the turbine is the same as that of the pump and equal to a value "*e*." Let E_2 equal the overall efficiency of the machine, and let η equal the thermal efficiency of the cycle. Let x be the work done by the turbine (in heat units) and y the work done by the pump (in heat units), and H the quantity of heat entering the furnace. Then—

$$\eta = \frac{x - y}{H}$$

and, as calculated for a steam turbine—

$$E_1 = \eta \cdot e = e \frac{x - y}{H}$$

but, with the gas turbine, however, this becomes—

$$E_2 = \frac{ex - y/e}{H} = \frac{e^2x - y}{eH}$$

for the negative work done *on* the pump increases while the positive work done by the turbine decreases.

Fig. 9 is a diagram showing the change in the value of the overall efficiency with the change in the value of the thermodynamic efficiency—this efficiency being assumed to be the same in the pump as in the turbine. Arbitrary values are taken of 3, 1, and 4, for x, y, and H, which give a value of ·5 for η, the thermal efficiency of the hypothetical system. The graph *oa* is drawn to the equation—

$$\eta = e \frac{x - y}{H}$$

on the steam engine basis, and the curve *ab* to the equation—

$$E_2 = \frac{ex - y/e}{H}$$

for the gas turbine.

With gas turbines of this type, however, it is always more efficient in practice to complete the cycle with isothermal

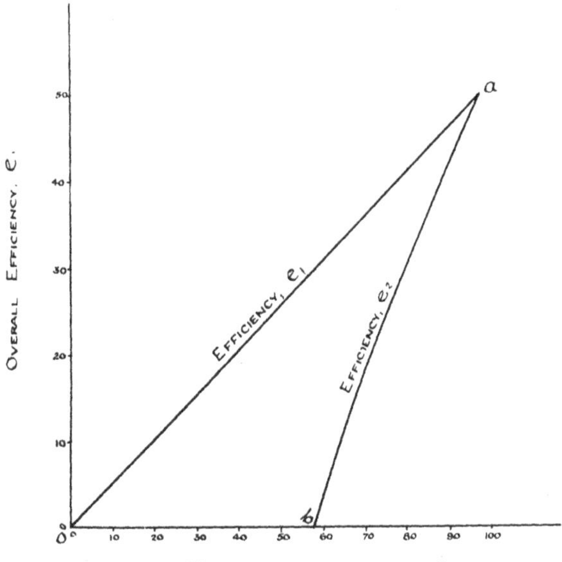

Fig. 9.—Diagram showing Change in Value of Overall Efficiency with Change in Thermodynamic Efficiency.

compression instead of with adiabatic compression, as in the Diesel engine. The thermal efficiency of the cycle is, of course, more efficient if the compression is performed *adiabatically* in all cases where the compressed gases can be delivered direct to the furnace, the heat of compression being all absorbed by the working fluid. The theoretical thermal efficiencies of the two systems can be compared as follows :—

For a cycle using adiabatic compression the thermal efficiency is wholly independent of the temperature of the working fluid, being only dependent on the ratio of compression. The case is exactly analogous to that of the Diesel oil engine, and the thermal efficiency, η_a, is given by the equation—

$$\eta_a = 1 - \frac{1}{r^{\gamma - 1}}$$

where r = ratio of compression.

In the case of isothermal compression the efficiency is worked out as follows :—

Let W be the work done by the turbine per lb. of fluid and let ω be the corresponding work done by the pump. Then the thermal efficiency, η_i,

$$= \frac{W - \omega}{C_p(\theta_1 - \theta_0)}$$

where θ_1 is the temperature of the working fluid before expansion, and θ_0 the lowest temperature of the cycle. Then—

$$\eta_i = \frac{C_p(\theta_1 - \theta_2) - v_0 p_0/J.\ (\log_e r)}{C_p(\theta_1 - \theta_0)}$$

where $v_0 p_0$ is the state of the gas before compression.

$$= \frac{C_p(\theta_1 - \theta_2) - C_p\,\theta_0 \log_e r^{\frac{\gamma - 1}{\gamma}}}{C_p(\theta_1 - \theta_0)}$$

$$= \frac{\theta_1(1 - 1/r^{\frac{\gamma - 1}{\gamma}}) - \theta_0 \log_e r^{\frac{\gamma - 1}{\gamma}}}{\theta_1 - \theta_0}$$

$$= \frac{\theta_1(r^{\frac{\gamma - 1}{\gamma}} - 1) - \theta_0 \log_e r^{\frac{\gamma - 1}{\gamma}}}{r(\theta_1 - \theta_0)}$$

It is thus seen that the efficiency in this case is dependent not only on the ratio of compression but also on the initial and final temperatures of the working fluid. The efficiency in this case only approaches the value of that obtained in the adiabatically compressed cycle when the ratio of θ_1 to θ_0 becomes very great. In the above equation, when θ_0 is zero (or when θ_1 is infinitely great), it will be seen that $\eta_i = \eta_a$.

Though there is this advantage in thermal efficiency of the adiabatic compression over the isothermal, the latter is more advantageous when the question of thermodynamic losses is taken into account. Assume, in a given case, that the compression ratio r is 10, that θ_1 is 1,000° C., and that θ_0 is the atmospheric temperature, viz., 15° C., then—

$$\eta_a = 1/10^{\gamma - 1} \quad (\gamma = 1{\cdot}38)$$
$$= {\cdot}58$$

while the efficiency with isothermal compression becomes—

$$\eta_i = {\cdot}50.$$

There is a gain here of 8 per cent. in the adiabatic cycle on the purely theoretical count. There is, really, however, a nett loss on the overall efficiency owing to the increased thermodynamic losses of the pump.

Assume the thermodynamic efficiency of the turbine to be 70 per cent. and that of the pump to be 70 per cent. Then, if W is the work done by the turbine and ω is the work done by the pump, and H is the quantity of heat put into the furnace per lb. of working fluid—

$$\text{Overall efficiency, E} = \frac{\text{W} \times 70 \text{ per cent.} - \omega/70 \text{ per cent.}}{\text{H}}$$

Consider the isothermal compression system.

Work done by turbine—

$$\text{W} = \frac{\gamma}{\gamma - 1} \cdot \frac{v_0 p_0}{\text{J.}} \cdot \frac{\theta_2}{\theta_0}\left(r^{\frac{\gamma-1}{\gamma}} - 1\right) \text{ T.U.}$$
$$= 147 \text{ T.U.}$$

Work done on pump in isothermal compression—

$$\omega_i = \frac{p_0 v_0}{\text{J.}} \log_\epsilon r \text{ T.U.}$$
$$= 43 \text{ T.U.}$$

Heat put into working fluid—

$$\text{H}_i = \text{C}_p\,(\theta_1 - \theta_0)$$
$$= 250 \text{ T.U.}$$

Therefore overall efficiency—

$$\text{E}_i = \frac{\text{W} \times 70 \text{ per cent.} - \omega \div 70 \text{ per cent.}}{\text{H}}$$
$$= \frac{103 - 61{\cdot}5}{250}$$
$$= 16{\cdot}6 \text{ per cent.}$$

Consider the adiabatic system.

Work done by turbine—

$$W = 147 \text{ T.U. (as above).}$$

Work done on pump—

$$\omega_a = \frac{\gamma}{\gamma - 1} \cdot \frac{v_0 p_0}{J} \cdot \left(r^{\frac{\gamma-1}{\gamma}} - 1\right) \text{ T.U.}$$
$$= 59 \text{ T.U.}$$

Heat put into working fluid—

$$H_a = C_p (\theta_1 - \theta_3) = 191$$

where θ_3 is the temperature after adiabatic compression in the pump $= \theta_0 r^{\frac{\gamma-1}{\gamma}}$.

Therefore overall efficiency—

$$E_a = \frac{W \times 70 \text{ per cent.} - \omega/70 \text{ per cent.}}{H}$$
$$= \frac{103 - 85}{191}$$
$$= 9{\cdot}5 \text{ per cent.}$$

It will be thus seen that although the *thermal* efficiency of the system using adiabatic compression is 16 per cent. greater than the *thermal* efficiency in the case of the isothermal compression ; yet the *overall* efficiency is 57 per cent. greater in the *latter* case than in the former.

The loss in this type of turbine due to the inefficiencies of the pump and turbine is shown graphically in Fig. 10. The diagram is a pressure-volume one, drawn for ten expansions through the turbine and an upper limit of temperature of 879° C., representing a volume expansion of four in the combustion chamber. AB represents the original volume of the working fluid ; this is expanded in the furnace to the volume AE ; AE = 4AB. The curve EF represents the adiabatic expansion through the turbine, drawn to the equation— pv^{γ} = constant. DM is the volume after expansion at the pressure p_2. The volume contracts to DL after cooling in the cooler or condenser. The pump compresses the fluid back to the original pressure p_1, isothermally along GB, or adiabatically along GH.

The curve *xm,* which bounds the shaded area on the left-hand side, shows the increase in the work done on the pump if the thermodynamic efficiency of the same is 70 per cent., the compression being isothermal, this work being now represented by the area AxmP. The curve yn bounding the darkened strip shows the loss in efficiency of the turbine, assuming the thermodynamic efficiency of the turbine to be the same as that of the pump. The work, then, done by the turbine is represented by the area AynP. The area $xynm$, then, is the net work available for use on the turbine shaft.

If the compression in the pump is adiabatic the work done

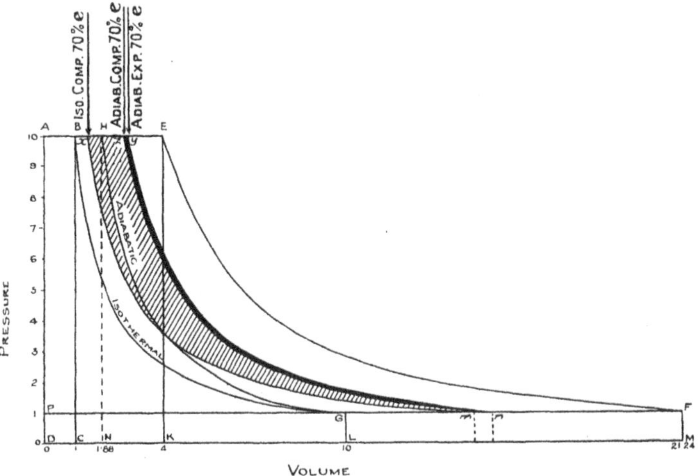

FIG. 10.—"PV" Diagram, showing Loss due to Inefficiencies in Turbine and Pump.

on the pump at 70 per cent. thermodynamic efficiency is represented by the area AzmP, the amount of work left over for external application after the turbine inefficiencies have been taken off being that represented by the blackened strip $zynm$.

The necessity of isothermal compression in this class of turbine will thus easily be realised. Of course, in those cases in which the compressed fluid is not returned to the furnace, but is rejected to the atmosphere, there is not even any gain on the thermal count.

Fig. 11 is a temperature-entropy diagram drawn for a gas turbine working on this cycle. The zero for entropy is taken at 0° C. The curve AB represents the change in entropy with

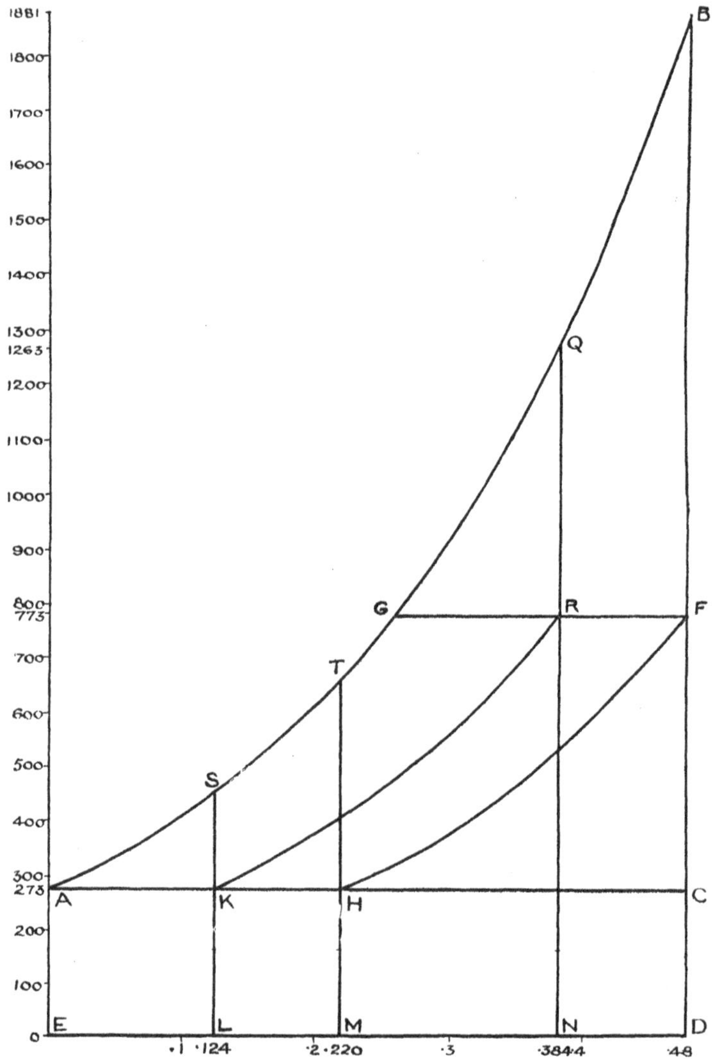

Fig. 11.—"$\theta\phi$" Diagram for Single-Fluid Constant-Pressure Gas Turbine.

the rise in temperature, heat being absorbed at constant pressure, drawn to the equation—

$$\phi = C_p \log \frac{\theta}{\theta_0}$$

on the supposition that the specific heat does not change with the temperature (*vide* Chapter VI.). Verticals, such as BF, represent adiabatic expansion ; the curve FH, heat rejection at constant pressure ; the horizontal HA, isothermal compression ; or the vertical HT, adiabatic compression.

For adiabatic compression the total heat put in is represented by the area underlying the curve TB, and the available work by the area TBFH. The area ATME is a measure of the work done on the pump in adiabatic compression ; the area HFDM, that of the heat rejected to the heat-sump.

With isothermal compression the area underlying the curve AB is the measure of the heat put into the working fluid ; the area ABFH, that of the available work ; and the area AHFDE, that of the total heat rejected in pump and cooler ; the area AKLE being the measure of the work done on the pump in isothermal compression. Further reference, of a quantitative nature, will be made to this diagram later, when the various types of constant-pressure gas turbines are being discussed.

Gas turbines working with a single fluid and taking in heat to the working fluid under constant pressure are broadly divisible into two classes. Those that are fired externally and those that are fired internally. Those, that is to say, in which a quantity of air (or other gas) is heated by passing through a number of tubes, heated by the external combustion of fuel, and those in which the fuel is burned with its requisite amount of air or oxygen, and the products of combustion expanded through the turbine.

THE EXTERNALLY-FIRED CONSTANT-PRESSURE GAS TURBINE.

The externally-fired type of gas turbine is the most elemental form in which the gas turbine (or, rather, the constant-pressure gas turbine) is capable of conception. From the purely practical point of view the system of external firing is wholly out of the question, as the difficulty of getting heat through metal walls to raise the temperature of so bad a conductor as air is enormous. They are interesting, however, from the theoretical point of view, as the starting-point from which all other constant-pressure gas turbines may be said to have been evolved. The system consists essentially of a heater of the tube or locomotive-boiler variety; a turbine, through which the working fluid expands and does work after receiving heat in the furnace; a cooler in which heat is abstracted from the fluid before it passes into the pump; and a pump in which the fluid is compressed isothermally to the initial pressure. In one class the cooler is dispensed with, the fluid being delivered after expansion into the atmosphere. The cooler may also be partly (theoretically it may be wholly) regenerative. The cycle may be "closed" or it may be "open"; that is to say, the same working fluid may be used repeatedly, or the turbine may exhaust into the atmosphere and the pump compress from the atmosphere, or the turbine take from the atmosphere and the pump compress to the atmosphere.

EXTERNAL FIRING: CLOSED CYCLE TYPE.

This type is shown in Fig. 12. F is the furnace, through which the working fluid receives heat on its way from the pump P to the turbine T; R is the regenerator; C is the cooler, in which heat is abstracted from the working fluid on its way from the turbine to the pump. With adiabatic compression in the pump the thermal efficiency of the cycle—

$$\eta = 1 - 1/r^{\gamma - 1}.$$

The system, however, is quite impracticable, owing to the impossibility of getting sufficient heat into the air through the tubes of the furnace with burning out the tubes in question. The state of things can be best seen by taking an actual example.

Let it be assumed that the tubes can be made to stand as high a temperature as 500° C. This is a sufficiently favourable assumption. Let the temperature difference necessary to convey the heat through the walls of the tubes be 100°. As a matter of fact, between air and air a figure very much greater than this is indicated. The temperature of the air entering the expanding nozzle is, then, 400° C. Let the working fluid be expanded down to the atmospheric temperature. In this case the cooler is unnecessary. Let the thermodynamic efficiency of the turbine be 80 per cent. It is reasonable to assume here an efficiency higher than that prevailing in ordinary turbine practice, as velocities here are small and the fluid used is non-condensible. Let the pump thermodynamic efficiency be 70 per cent.: it is assumed that a rotary or turbine air pump is employed. A piston air pump would give a higher efficiency. but it would have to be driven from the turbine by

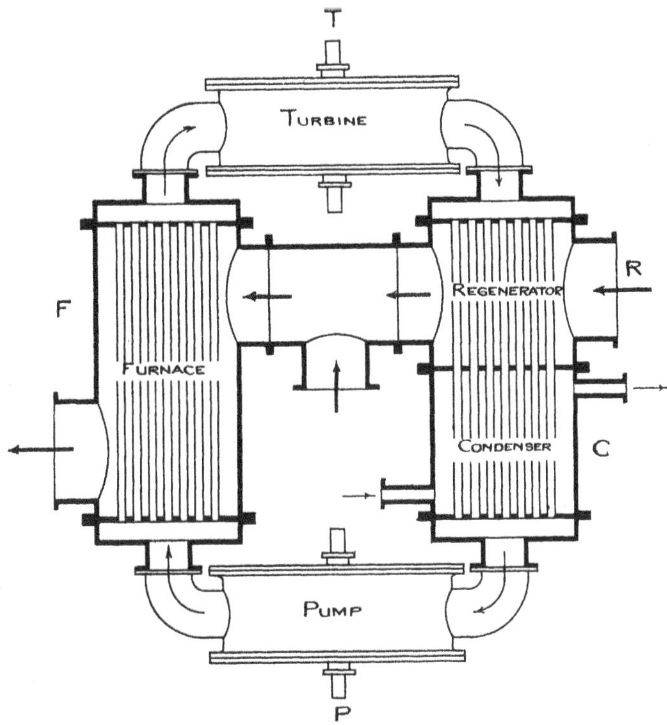

FIG. 12.—Constant-Pressure Gas Turbine. External Firing. Closed Cycle. Partial Regeneration.

means of gearing, which would bring its efficiency down to the value quoted.

Consider the work done in the turbine per lb. of air. Then—

$$W = \frac{\gamma}{\gamma - 1} v_0 p_0 \left\{ \frac{\theta_1 - \theta_0}{\theta_0} \right\} \text{ ft.-lbs.}$$
$$= 96{\cdot}25 \text{ T.U.}$$

The work done on the pump—

$$\omega = \frac{\gamma}{\gamma - 1} p_0 v_0 \log \frac{\theta_1}{\theta_0}$$
$$= 57{\cdot}80 \text{ T.U.}$$

Therefore thermal efficiency of cycle—

$$\eta = \frac{96{\cdot}25 - 57{\cdot}80}{96{\cdot}25 + h}$$

where h = heat carried away per lb. of gas from furnace.

At 100° difference between inside and outside of tubes—

$$h = C_p\, 100 = 25 \text{ T.U.}$$
$$\therefore \quad \eta = \frac{96{\cdot}25 - 57{\cdot}80}{131{\cdot}25} = 29 \text{ per cent.}$$

With the thermodynamic efficiency of the turbine at 80 per cent., and that of the pump at 70 per cent., the work done by the turbine—

$$W = 96{\cdot}25 \times 80 \text{ per cent.} = 77 \text{ T.U.}$$

and the work done on the pump—

$$\omega = 57{\cdot}8 \times 1/{\cdot}7 = 83 \text{ T.U.}$$

Therefore the work done per lb. of air passing through the system by the gas turbine equals −6 T.U. Or, in other words, some 8,000 ft.-lbs. will have to be exerted *on* the turbine per lb. of working fluid passing in order to get the machine to turn round at all.

It is clear, therefore, that external firing is impossible in practice owing to the low temperature that it is possible to get into the working fluid.

EXTERNAL FIRING : OPEN CYCLE.

This type of gas turbine differs only from the preceding type in the fact that the same working fluid is not used continuously in the system. Were it possible to produce an external-fired gas turbine that would admit of high temperatures being used, the closed cycle would have this advantage over the open cycle that the mean pressure of the air in the turbine could be purely arbitrary, and consequently a large quantity of work could be produced with the minimum of fluid present.

The open cycle, however, while being equally impossible in

practice as the closed cycle, owing to the external firing, has nevertheless a peculiar advantage of its own over all other types of gas turbines. This is the ability to achieve *complete regeneration*. A turbine of this type is shown diagrammatically in Fig. 13. T is the turbine, F the furnace, and P the pump.

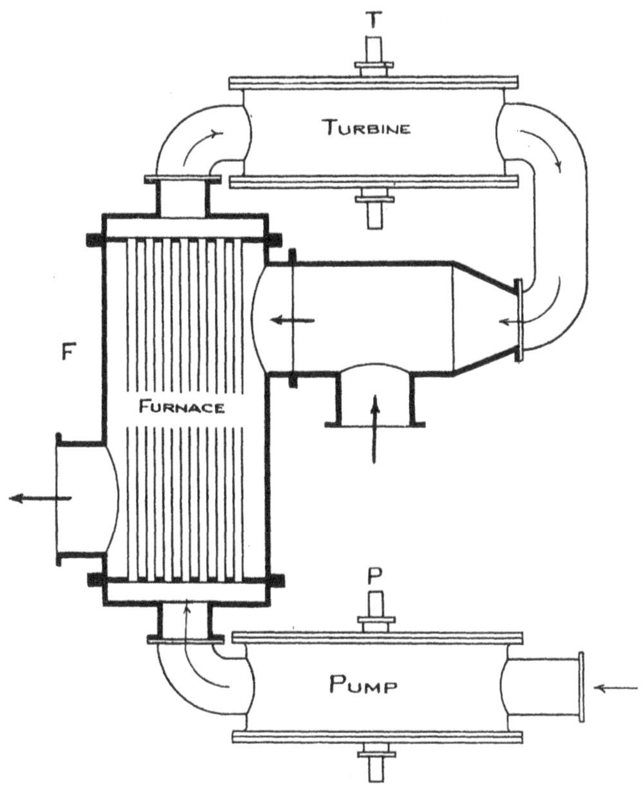

FIG. 13.—Constant-Pressure Gas Turbine. External Firing. Open Cycle. Complete Regeneration.

The expansion and compression take place *entirely* above the atmosphere. The air, after expansion in the turbine, is delivered to the furnace and burnt with the addition of fuel. Consequently all the heat that is in the exhaust from the turbine is recovered in the furnace. The thermal efficiency of the system is, therefore, 100 per cent. As, however, it suffers from exactly the same practical defects as the closed cycle type, the highest practical overall efficiency is a minus quantity, being represented by about 8,000 ft.-lbs. per lb. of air passing through the turbine, to be applied externally to turn the machine round.

Fig. 14 shows a similar arrangement to that in Fig. 13, but with the expansion of the gas wholly below the atmosphere.

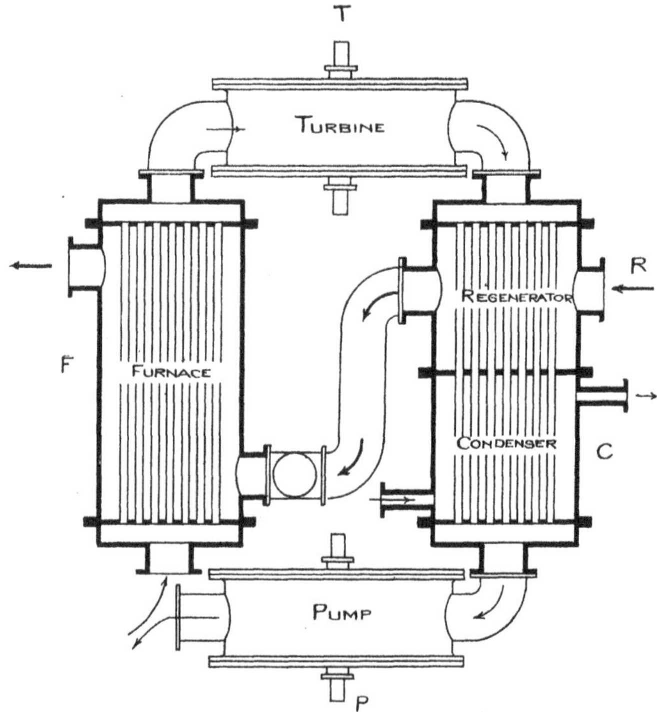

Fig. 14.—Constant-Pressure Gas Turbine. External Firing. Open Cycle. Sub-atmospheric.

In this case only partial regeneration is possible. The figure is sufficiently explanatory. The thick line shows the course of the working fluid, the thin line the course of the externally applied heating agent. This convention is kept throughout the chapter.

All the gas turbines dealt with so far are of theoretical interest only. External heating is wholly impracticable ; but it may be borne in mind that one type in this class of turbine presents the highest thermal efficiency that it is possible to obtain in any heat engine, viz., a ratio of 100 per cent. between the heat put into the system and the work available from the cycle.

It may not, perhaps, be altogether out of place here to lodge something in the nature of a protest against the manner in which the thermal efficiency of a cycle is occasionally represented. It is not uncommon to find two cycles compared together on the grounds of thermal efficiency. This comparison is often not only inaccurate, but positively fallacious, and may have no value as an index to the relative value of the two systems. That a system has a higher thermal efficiency than another system is no indication *whatsoever* that the overall efficiency of the former is greater, or even *likely* to be greater, than the latter. For example, the thermal efficiency of the gas turbine of the type shown in Fig. 13 is, as above stated, 100 per cent. The thermal efficiency of a steam engine working between 150 lbs. and 2 lbs. absolute without superheat, is approximately 27 per cent. If the thermodynamic efficiency (or " ratio efficiency ") of the steam engine be 70 per cent., then the overall efficiency of the engine is 27 per cent. × 70 per cent., viz., 19 per cent. If the thermodynamic efficiencies of the turbine and pump in the gas turbine above alluded to are 70 per cent., though the thermal efficiency is more than three times that of the steam engine, the overall efficiency falls below zero. It can truly be said that in any *given* cycle, if the thermal efficiency is raised (the thermodynamic efficiencies being constant), then the overall efficiency is raised. It *cannot* truly be said that if in any one cycle the thermal efficiency is greater than in any one other, then (the thermodynamic efficiencies being equal) the overall efficiency in the former is greater than the overall efficiency in the latter. To argue in favour of any cycle on the grounds of " high thermal efficiency " alone is precisely in the same category as recommending a workman on the grounds " that he can use his tools skilfully *when he can be got to work !* " The main question is " Can he be made to work ? " As the output is the only matter of any moment to anybody, it is better to have a second-class workman who will work, than a first-class workman who will not. Such obvious truisms as these would not be reiterated were it not for the obsession amongst not a few scientists to regard thermal efficiency as a kind of cardinal virtue, wholly excellent in itself, and rather tending to ignore the fact that there is only one cardinal virtue in all questions of prime movers, and that is "*The minimum expenditure of money for the maximum return of power.*"

THE INTERNALLY-FIRED, CONSTANT-PRESSURE, SINGLE-FLUID TURBINE.

A quantity of gas and air is burned in a suitable vessel and the heated products of combustion expanded through a turbine, a pump being used to maintain the necessary difference in pressure, and a cooler being employed to abstract the heat from the effluent gases of the turbine in the case of expansion below the atmosphere.

In cases where the compression of the working fluid takes place wholly above the atmosphere there is no need for a cooler, the turbine simply exhausting into the atmosphere and the pump compressing from the atmosphere. An arrangement of this kind is shown in Fig. 15. The pump P compresses the air and delivers it through the regenerator R to the furnace F, where it is burnt in conjunction with the fuel. If the fuel is gaseous it has to be compressed first to the requisite pressure by a rotary pump (similar to P) not shown in the diagram; if the fuel is a liquid (*i.e.*, oil) it can more conveniently be compressed by a piston pump (also not shown in the diagram). The exhaust gases from the turbine T part with some of their heat in the regenerator R to the air going into the furnace. A partial regeneration is thus effected; it is impossible to effect a complete regeneration owing to the temperature difference between the inner and outer walls of the regenerator tubes necessary to ensure the transfer of heat.

Fig. 16 represents a similar type of turbine arranged for expansion wholly below the atmosphere. The fuel is burnt in the furnace F at atmospheric temperature; the products of combustion are expanded through the turbine T; a part of the remaining heat that they still contain is abstracted by the entrant air in the regenerator R; the remainder by water-cooling in the cooler or condenser C. The waste gases are restored to the atmospheric pressure (isothermally) by the pump P.

Theoretically, of course, the presence of the water-cooler could be dispensed with, but in practice it is not possible to cool the gases sufficiently by air alone, and were further cooling

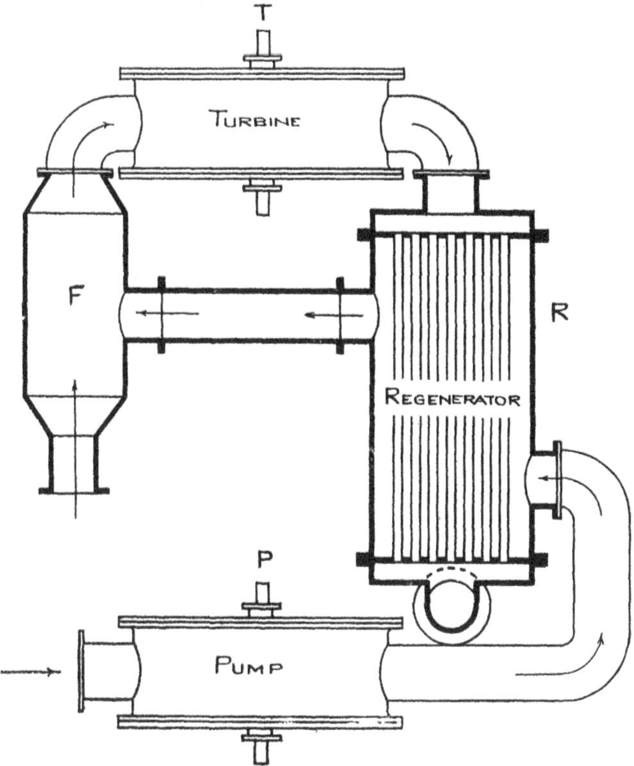

Fig. 15.—Constant-Pressure Gas Turbine. Internal Firing. Super-atmospheric.

by water to be done away with, the work of the pump would be unduly increased. Although the addition of the condenser of necessity complicates the turbine, still the turbine in which the expansion takes place wholly below the atmosphere has the great advantage over that in which the expansion is above the atmosphere, owing to the fact that combustion can take place at the atmospheric pressure. Continuous burning under super-atmospheric pressures has always presented practical difficulties; where it can be avoided it is advisable to do without it. There is also a gain in the fact that all rotary pumps give a higher thermodynamic efficiency when the mean density of the fluid in which they operate is lower, owing

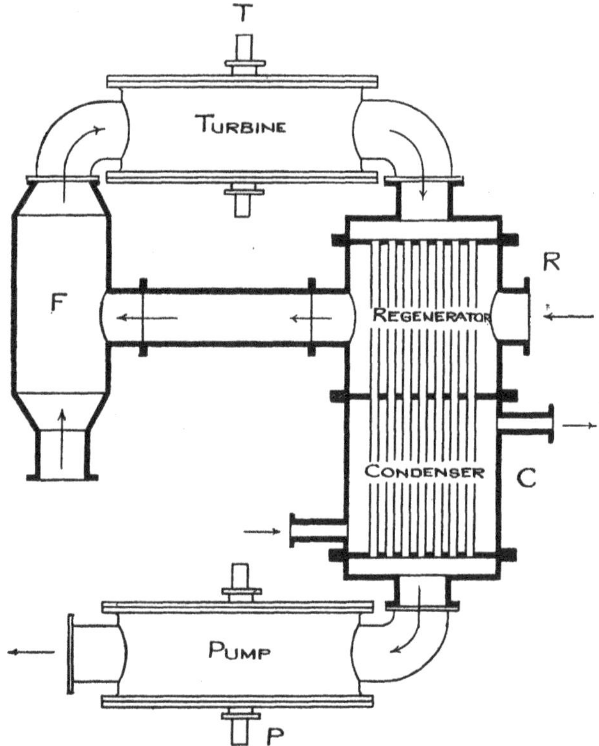

FIG. 16.—Constant-Pressure Gas Turbine. Internal Firing. Sub-atmospheric.

to the decrease of skin friction of the rotating parts, a factor that decreases with the density of the medium and which is a very considerable factor with high peripheral speeds (*vide* Chapter VII.).

The theoretical efficiency of the cycle is the same in either case. The great advantage that the internal combustion turbine possesses over the external combustion turbine is in the capability of the use of very much higher temperatures in the furnace. In the former case the furnace can be lined with a refractory material and practically any temperature can be used. The temperature on the turbine blades is not the temperature in the furnace before expansion, as the expansion in the expanding nozzle is adiabatic and accompanied by the corresponding drop in temperature.

There are only two factors that determine (on the theoretical count) the maximum temperature that it is possible to employ in the combustion chamber. These two factors are:—

(1) The maximum temperature that is permissible upon the turbine blades.

(2) The maximum peripheral velocity that it is possible to give to the turbine wheel.

The first settles the temperature of the working fluid after expansion through the first nozzle. Let this maximum permissible temperature be 500° C. It now remains to fix the upper limit of temperature of the cycle by the amount of energy that can be effectively absorbed by the first turbine wheel.

De Laval turbines have been run with peripheral velocities approaching 1,400 feet per second. This is, however, phenomenal, and may be ruled out of court; it is only applicable to geared turbines. Stumph turbines have been built to run at peripheral speeds as high as 1,200 feet per second, but this again errs on the side of excessive speed. For gas turbines (in which much higher temperatures are dealt with than in steam turbine practice) the maximum limit of peripheral velocity may be taken at 1,000 feet per second.

If a "double wheel," that is, a wheel with two rings of blades, as used in the Curtis turbine, be employed, the ratio of the speed of the blades to the velocity of the impinging gas may be taken as 1 to 5. This is the usual factor in steam turbine practice for wheels of this nature. The velocity, therefore, of the working fluid will be 5,000 feet per second.

Now the work done per lb. of working fluid, that is, the amount of work that this velocity represents—

$$W = \frac{v^2}{2gJ} = 277 \text{ T.U.}$$

where g is the acceleration of gravity and J is the mechanical equivalent of heat, and $v =$ the velocity of the gas $= 5{,}000$ ft./sec.

Now W, the work done in the first expansion, also equals—

$$\frac{\gamma \cdot v_0 p_0}{(\gamma - 1)\,\theta_0 \mathrm{J}} (\theta_1 - \theta_2) = 277 \text{ T.U.}$$

Taking γ to be 1·38 for the products of combustion, and filling in the values of the other constants, we get—

$$\theta_1 - \theta_2 = \frac{277}{\mathrm{C}_p} \text{ T.U.}$$

$$(\mathrm{C}_p = \cdot 25)$$

$$\theta_1 - \theta_2 = 1108°.$$

Now $\theta_2 = 773°$ A. Therefore the upper limit of temperature, θ_1, equals 1881° A, or 1,608° C.

We are now in a position to ascertain the maximum efficiency that a turbine of this class can be expected to give. It may first be seen how many expansions this drop in temperature of 1,108° represents. The work done in this expansion—

$$\mathrm{W} = \mathrm{C}_p\,\theta_2\,(r^{\frac{\gamma-1}{\gamma}} - 1)$$

Therefore

$$\theta_1 - \theta_2 = \theta_2(r^{\frac{\gamma-1}{\gamma}} - 1).$$

This gives r, the expansion ratio, equal to 25·2.

As a matter of fact, no turbine compressors have ever been made to pump to so high a compression as this ; the number of elements required (at reasonable peripheral speeds) would be excessive. The matter need not be gone into here, it will be considered fully when the question of the pump is dealt with in Chapter VII.

If reference is made to Fig. 11, p. 20, the temperature-entropy diagram for this expansion will be seen. AB represents the reception of heat by the gas at constant pressure to the temperature of 1881° A, at B ; BF represents the adiabatic expansion to the temperature 773° A ; FH, the rejection of heat at constant pressure ; and KA, the isothermal compression of the fluid to the original pressure. The thermal efficiency of the cycle is then represented by the area ABFH over the area ABDE.

The thermal efficiency of the cycle is found as follows :—

The work done by the turbine—

$$W = C_p(\theta_1 - \theta_2)$$
$$= 277 \text{ T.U. (as above).}$$

The work done on the pump (isothermal compression)—

$$\omega = \frac{v_0 p_0}{J} \log_e r$$
$$(r = 25{\cdot}2)$$
$$= 60{\cdot}4 \text{ T.U.}$$

Heat put in to working fluid per lb. of same—

$$H = C_p(\theta_1 - \theta_0)$$
$$= 398{\cdot}25 \text{ T.U.}$$

Then the thermal efficiency of the cycle—

$$\eta = \frac{W - w}{H}$$
$$\doteq 54{\cdot}4 \text{ per cent.}$$

Here, however, no account has been taken of regeneration.

Assume that the temperature difference necessary to convey the heat through the walls of the tubes of the regenerator to be 200° C. This is here assumed at a more probable value than the rather sanguine figure assumed earlier in this chapter as regards the externally-fired gas turbine.

Then the heat rejected in the exhaust from the turbine is—

$$h = C_p(200) = 50 \text{ T.U.}$$

Therefore total heat put into system equals—

$$W + h = 327 \text{ T.U.}$$

Then thermal efficiency—

$$\eta = \frac{W - w}{W + h} = \frac{216{\cdot}6}{327} = 66{\cdot}2 \text{ per cent.}$$

Let the thermodynamic efficiency of the turbine be 70 per cent. and that of the pump be 70 per cent.; then the overall efficiency—

$$E = \frac{W \times 70 \text{ per cent.} - \omega/70 \text{ per cent.}}{W + h}$$
$$= 36 \text{ per cent.}$$

It will thus be seen that it is possible to get an overall

efficiency higher than that obtained even with the Diesel oil engine.

The above is the highest overall efficiency obtainable in a single-fluid gas turbine, taking in heat at constant pressure, within the limits of temperature, pressure, speed, and thermodynamic efficiency that can be admitted upon the most favourable survey of contemporary practice in cognate cases. It has been thought advisable to devote a special chapter to the analysis of these practical limitations, as the whole problem of the gas turbine may be said to reside wholly in this matter. The reader is referred to Chapter VIII., where the question of peripheral speed, blade temperature, thermodynamic efficiencies, compression ratios, limits of regeneration, etc., are discussed in detail; for the reasons of assigning the limits that have been adopted in the foregoing calculation of efficiency, this chapter, with those on the rotary pump and accessories, must be consulted.

The above state of affairs is, as has been already stated, of the most favourable kind. It would be not uninteresting to work out the overall efficiency of the above cycle under a condition of practical limitations more consonant with common practice.

Let the lower limit of temperature (that is, the temperature on the turbine blades) be, as before, 500° C. Let the peripheral speed of the turbine rotor be 700 ft./sec.; a value more approaching ordinary turbine practice than that of 1,000 ft./sec. There is some loss in actual expanding nozzles owing to the friction of the expanding fluid against the walls of the nozzle. Let the nozzle efficiency be 90 per cent. This will mean that the working fluid becomes "super-heated," and does not drop to the full limit of the temperature drop as ascertained by the calculation from the kinetic energy equation. Let the ratio of the speed of the blade to the speed of the fluid be, as before (a "double" wheel being used), 1 to 5. Then the velocity of the gas is 3,500 feet per second, and W the work done, equals—

$$\frac{v^2}{2gJ} \text{ T.U.}$$
$$= 136 \text{ T.U.}$$

Now

$$W = \cdot 25\,(\theta_1 - \theta_2).$$

Therefore the theoretical temperature drop is 544°. Actual temperature drop at 90 per cent. nozzle efficiency is 490°. Also

$$W = C_p \theta_2 (r^{\frac{\gamma - 1}{\gamma}} - 1)$$

therefore $r = 6{\cdot}9$ (approximately).

The work done in isothermal compression on the pump—

$$\begin{aligned} \omega &= p_0 v_0 \log_e 6{\cdot}9 \\ &= 36 \text{ T.U. (approximately).} \end{aligned}$$

Let 200° difference be required (as before) in regenerator ; then heat put into system—

$$H = 136 + 50 = 186 \text{ T.U.}$$

Therefore the thermal efficiency—

$$\eta = \frac{W - w}{W + h} = 53{\cdot}8 \text{ per cent.}$$

If the thermodynamic efficiency of the pump and turbine each equal 70 per cent., then the overall efficiency—

$$E = 23{\cdot}6 \text{ per cent.}$$

It will thus be seen that under what may be considered as normal conditions of temperature and peripheral velocity, the overall efficiency of a gas turbine of this class exceeds that of the steam turbine, approaches that of the gas engine, and falls short of the Diesel oil engine by some 10 per cent.

It is plain that a turbine of the foregoing kind may be made to expand the working fluid both above and below the atmospheric pressure ; the peculiar advantage thus afforded being that of increased range of pressure without increasing the mean density of the fluid. Thus with a maximum pressure of 60 lbs. absolute and a minimum pressure of 3 lbs. absolute, an expansion ratio of 20 is secured. If the final pressure were that of the atmosphere, the initial pressure would have to be 300 lbs. to attain the same number of expansions.

A turbine system of this nature in which the expansion is both sub- and super- atmospheric is shown diagrammatically in Fig. 17.

The compression pump P_1 delivers air through the regenerator R, at some pressure above that of the atmosphere, into the furnace F ; it is here burnt in conjunction with the necessary quantity of fuel (compressed to the requisite pressure by a separate pump, not shown in the diagram) and the products of combustion delivered to the turbine T, where they are adiabatically expanded to the lower limit of pressure. They

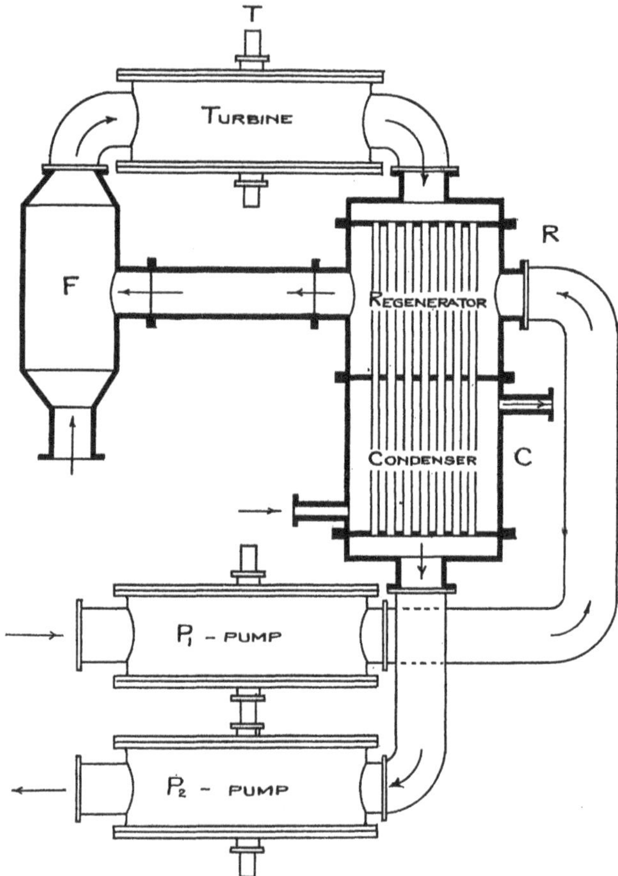

FIG. 17.—Constant-Pressure Gas Turbine. Internal Firing. Sub- and Super- atmospheric.

then pass into the regenerator R, where they part with a portion of their heat to the air feeding the furnace; the remainder of the heat is then abstracted by the water in the cooler C, and the gas is then compressed by the pump P_2 to the atmospheric pressure, and delivered into the atmosphere.

This is the first type of gas turbine yet considered that can be considered in any sense as a practical machine—that can be looked to by any manner of means as a competitor in the future to other heat engines. As has already been seen, the overall efficiency is only dependent on the limits of temperature, speed,

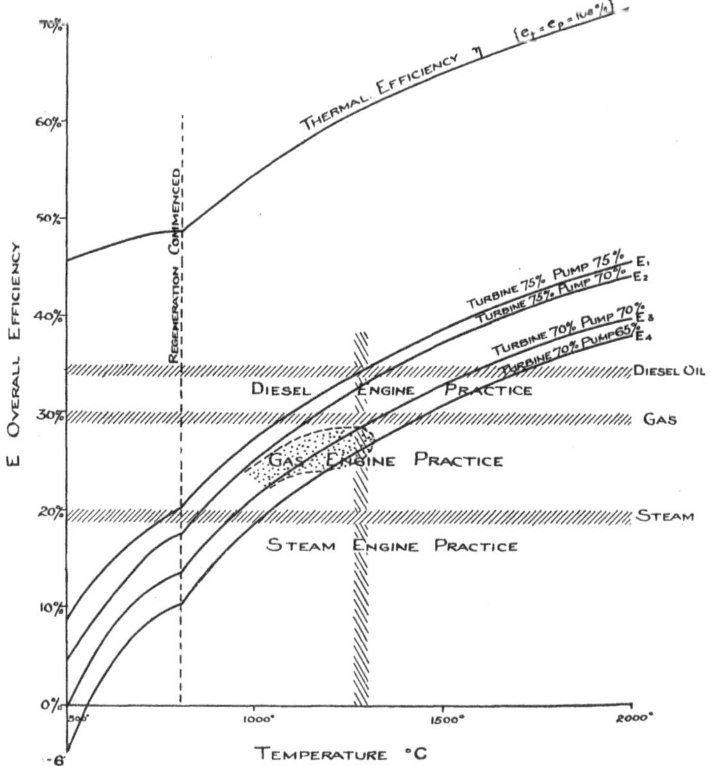

FIG. 18.—Efficiency-Temperature Diagram for Constant-Pressure, Single-Fluid, Internally Fired Gas Turbine.

and thermodynamic efficiency of the turbine and of the pump, which are constants fixed by practice. Any advance made in the durability of the steel used to make the blades or turbine rotor ; any decrease in the losses inherent in the turbine or pump ; allow of an increase in economy, whether by increase in initial temperature or otherwise. The increase of the overall efficiency with the rise in initial temperature is shown in Fig. 18. The diagram shows the variation of overall efficiency with temperature; the former as ordinates, the latter as abscissæ. Graphs are drawn for varying values of thermodynamic pump and turbine efficiency. The shaded areas represent the *loci* of efficiencies for the steam engine, the gas engine, and the Diesel.

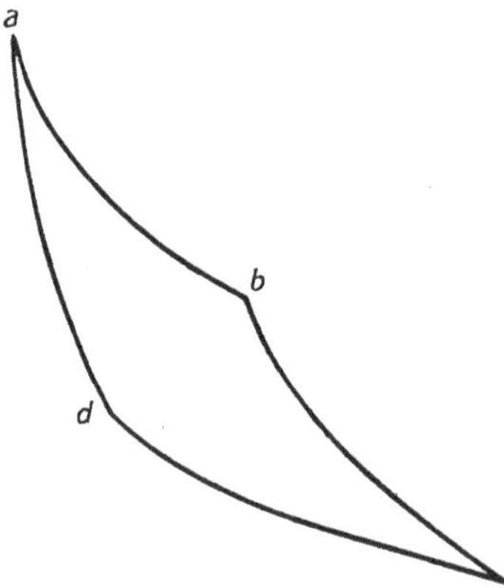

FIG. 19.—"PV" Diagram for Carnot Cycle.

The vertical shaded line represents the practical limit of initial temperature—about 1,300° C. The dotted area represents the *locus* of efficiencies for the gas turbine likely to be obtained in practice.

The sudden change in the slope of the curves at the vertical dotted line is caused by the fact that to the right of this temperature value, regeneration is calculated for; below this temperature, regeneration is not practically possible and is neglected. All the efficiency curves are drawn for twenty compressions in the pump.

These are the possible types of single-fluid gas turbines which take in heat at constant pressure. In all cases, the gas undergoes adiabatic expansion in the turbine. It is possible to conceive a turbine in which, by a judicious system of interheating between the elements, a state of affairs approximating to isothermal expansion may be realised. This condition is possible of conception; it is not possible of fulfilment, but cycles of this nature may be glanced at briefly. They are of some theoretic interest as comprising the "ideal heat engine." Two cycles are to be considered: the Carnot cycle and the

FIG. 20.—"$\Theta\Phi$" Diagram for Carnot Cycle.

Ericsson cycle. Neither of the cycles, of course, come into the category of taking in heat at constant pressure, though the latter does so (from the regenerator—or elsewhere) partially.

THE CARNOT CYCLE.

This well-known cycle consists of four operations : isothermal expansion ; adiabatic expansion ; isothermal compression ; and adiabatic compression. The pressure-volume diagram for the cycle is shown in Fig. 19. *ab* represents the isothermal expansion ; *bc* the adiabatic expansion ; *cd* the isothermal compression ; and *da* the adiabatic compression back to the initial conditions of temperature, pressure, and volume.

The entropy-temperature diagram for this cycle, of course,

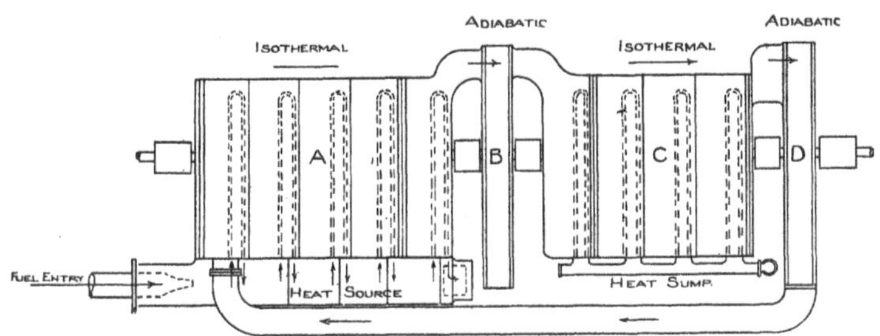

Fig. 21.—Diagrammatic Representation of Carnot Gas Turbine.

is represented by a rectangle, as shown in Fig. 20. The general arrangement of such a turbine is shown in Fig. 21. A is a turbine in which a condition approximating to that of isothermal expansion is approached by a system of reheating between the elements. Proposals of this nature have from time to time been made. A scheme for reheating with steam turbines, so as to secure isothermal instead of adiabatic expansion through the turbine, has been worked out in detail by Ferranti (*vide* Patent No. 24,781, 1902). After the working fluid has been expanded isothermally through A, it enters the turbine B, where it undergoes adiabatic expansion. Isothermal compression is effected by the rotary pump C, inter-cooling being achieved in the usual manner between the elements of the pump. Finally, the fluid is compressed adiabatically by the pump D, and delivered to the turbine A. It will be noticed that the cycle is a closed one, the same working fluid being used over and over again. The thermal efficiency is, of course—

$$\frac{\theta_1 - \theta_2}{\theta_1},$$

where θ_1 is the temperature at which heat is absorbed, and θ_2 is that at which heat is rejected. The system suffers precisely the same disadvantages as a practical machine as any externally-heated gas turbine, the initial temperature being limited to some 500° C. Placing the lower limit of temperature at that of the atmosphere, this gives a thermal efficiency of—

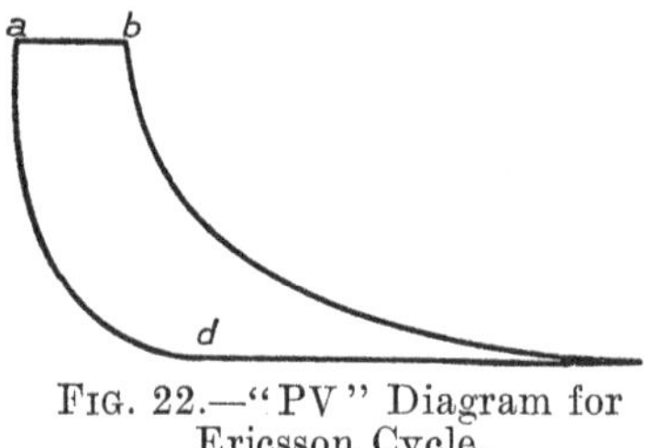

FIG. 22.—"PV" Diagram for Ericsson Cycle.

$$\frac{773 - 288}{773} = \cdot 615.$$

In actual practice the thermodynamic inefficiencies of turbine and pump would render the system abortive.

THE ERICSSON CYCLE.

This is the thermodynamic cycle adopted in the Ericsson air engine. It is similar to the Stirling cycle, except that heat is added and abstracted from the fluid at constant *pressure* instead of at constant volume. The pressure-volume diagram for the cycle is shown in Fig. 22. Heat is taken in at constant pressure along the line *ab*; isothermal expansion occurs along the curve *bc*; heat is rejected along *cd*, at constant pressure; isothermal compression takes place from *d* back to *a*. Fig. 23 is an entropy-temperature diagram for the same cycle; it is made up of two lines of heat-absorption at constant pressure, and two isothermals.

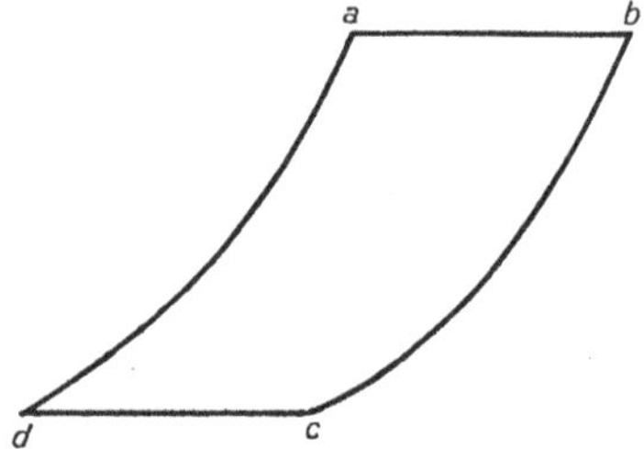

FIG. 23.—"ΘΦ" Diagram for Ericsson Cycle.

The general arrangement of such a turbine is shown in Fig. 24.

A represents the turbine, working isothermally by inter-elemental heating; B, the isothermal compression pump. C is the heat-sump; this may be partially regenerative.

If complete regeneration were to be secured the efficiency of this cycle becomes the same as that of the ideal heat engine, viz.—

$$\frac{\theta_1 - \theta_2}{\theta_1}.$$

The same objections apply to this system as to the foregoing:

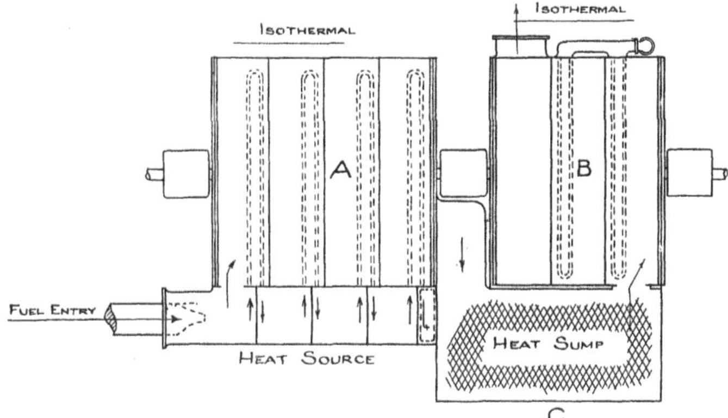

FIG. 24.—Diagrammatic Representation of Ericsson Gas Turbine.

isothermal expansion, indeed, is not a condition practically attainable with gas turbines.

It may be noted that in this cycle, unlike the foregoing one, the cycle is "open," a fresh supply of working fluid entering the furnace continually.

CHAPTER III. MIXED-FLUID TURBINES

So far, only those gas turbines using one fluid for the expansible medium have been considered. Constant-pressure gas turbines, however, have been made in which the working fluid consists of a mixture of the products of combustion from gas (or oil) and air, together with steam, produced by the injection of water into, or circulation around, the combustion chamber. The experimental turbine of Armengaud and Lemale worked (or failed to work) on this plan. Turbines of this type would seem to give a better chance of economy comparable with other prime movers than any other form of gas turbine, working within the present limits of temperature under practical working conditions. A reservation must be made with respect to the gas explosion turbine—the gas turbine taking in heat at constant volume—which may or may not produce the high efficiencies which at present it does *not* seem inclined to do, and which has always been, and is still, the " dark horse " of gas turbine research. The constant-volume gas turbine will be discussed fully later. The steam-and-gas turbine can be looked at from two points of view, according as to whether it is regarded as an evolution product of the steam turbine or of the constant-pressure gas turbine. It is, of course, possible to conceive a constant-volume gas turbine operating with more than one fluid, but such a system would have no advantage over the single-fluid explosion turbine, and a great many disadvantages over every other form of combustion turbine conceivable ; such need not be discussed here. The steam-and-gas turbine is essentially a constant-pressure turbine. It may be considered as evolved from the steam turbine, in the following way.

In any steam turbine and boiler installation it is well known and much regretted that a very considerable portion of the heat of combustion is lost in the boiler. This loss seldom falls below 20 per cent., even on test, with the best boilers, and often exceeds 40 per cent. with boilers of inferior design, under normal conditions of working. This is a very large loss, and from time to time proposals have been made to burn fuel under pressure and utilise the products of combustion and the steam in the engine cylinder, a piston pump being provided for maintaining the pressure of the air in the furnace.

This is essentially what is done in the steam-and-gas turbine. It merely consists of a steam turbine in which the products of combustion used in raising the steam, instead of going into the flue, are used in the turbine, together with the steam, the air being removed by means of a rotary pump. This is one way of looking at the steam-and-gas turbine. It may, however, be regarded from a different standpoint.

In a single-fluid gas turbine of the type that has been considered in the last chapter, the temperatures are apt to become too high for practical working, and water may be injected into the combustion chamber to reduce the temperature of the flame. In other words, the steam-and-gas turbine may be regarded as a constant-pressure gas turbine in which steam is used as a dilutant. The gradation from the single-fluid gas turbine, using steam as a dilutant to the mixed-fluid steam-and-gas turbine using the products of combustion from the boiler in the turbine with the steam, is simply one of quantity; not one of kind. As a matter of fact, the mixed-fluid turbine is, within the practical limit of temperature, more efficient than the single-fluid turbine, and the gas turbine using the maximum possible percentage of steam in the working fluid is more efficient than that using any less quantity of steam.

This maximum quantity of steam, which can be used as a dilutant to the working fluid, is fixed by the initial temperature that is to be used in the combustion chamber. Suppose this temperature to be 1,000° C. One lb. of combustion mixture composed of Mond gas and air in the requisite proportions for complete combustion (approximately equal volumes) gives out on combustion 550 T.U. Let the mixture be initially at atmospheric temperature. If the specific heat at constant pressure, C_p, of the mixture be taken at ·25, this would produce (theoretically) a temperature of 2,215° C.; provided there was no dilution, no radiation losses and no change in specific heat with rise in temperature, all of which contingencies, however, take place in actual fact.

It is not desirable here to take into account these last two variable factors; they will be dealt with in detail later. In the general consideration of the theory of the mixed-fluid constant-pressure turbine, the specific heats of the working fluids will be regarded as constant, and the combustion chamber will be considered to be so clothed with non-conducting material as to render any flow of heat to be reduced to a negligible quantity. It has been stated above that 1 lb. of combustion mixture of Mond gas and air gives out on combustion 550 T.U., and that the theoretical temperature produced would be 2,215° C., taking the specific heat of the combustion mixture to be ·25. As it is impossible to use temperatures of this order in a gas turbine, it is necessary to reduce the temperature by dilution. Suppose the practical limit of temperature at the entry to the expanded nozzle be 1,000° C. (*vide supra*). In the single-fluid type of turbine considered in the last chapter the dilution was effected by the admission of excess of air. If x be the additional weight of air to be added to 1 lb. of combustion mixture, then—

$$550 = 985\,(\cdot 25 + \cdot 24x),$$

the specific heat of air being taken at ·24. This gives a value for x of 1·08 lbs.

In the steam-and-gas turbine, however, this dilution has to be effected by the addition of steam only. Let the steam be at atmospheric pressure in the combustion chamber. Then the total heat of 1 lb. of steam, measured from 15° C., is 624 T.U. The super-heat to be added is—

$$(1{,}000 - 100)\,\cdot 5 = 450 \text{ T.U.}$$

and the total heat required to raise 1 lb. of steam from atmospheric temperature to the required conditions is 1,074 T.U. If y is the quantity of dilutant steam to be added, then—

$$550 = 985 \times \cdot 25 + 1{,}074y.$$

Therefore

$$y = \cdot 284 \text{ lbs.}$$

A much larger proportion of steam per lb. of combustion mixture may, in point of fact, be added without reducing the initial temperature of the working fluid, by means of regeneration. This matter will be considered fully when the question of regeneration is discussed.

It is, of course, obvious that any proportion of steam up to the limiting value may be mixed with the products of combustion, the excess dilution being effected by means of air ; it is also clear that the lower the initial temperature required the greater is the proportion of steam that may be introduced into the system.

It may not unfairly be asked, " Why introduce steam as a dilutant in preference to air ? " Steam is admittedly an inferior working fluid to air on the count of thermal efficiency. The loss in the former owing to the latent heat of evaporation is very heavy and steam will only give 25 per cent. efficiency, when air alone under similar conditions will achieve a thermal efficiency of something of the order of 50 per cent. The advantage accruing from the introduction of steam into the system lies in the reduction of the negative work in proportion to the positive work ; in other words, the " thermodynamic efficiency " is considerably increased ; and, between the limits of temperature possible in actual practice, the thermodynamic efficiency of the system is increased so much more than the thermal efficiency of the cycle is decreased, that the balance in overall efficiency lies with the steam-and-air turbine and not with the single-fluid constant-pressure turbine. There are also certain practical advantages of a minor nature coincident to the use of steam. This gain in economy due to the addition of steam is only extant below a fixed temperature limit, and below a fixed value of the thermodynamic efficiencies of the turbine and pump. But these limiting values of temperature and efficiency are not accessible under the conditions of contemporary practice.

Mixed-fluid turbines may be divided broadly into two classes: those in which the steam and air are actually mixed together and expanded through the same turbine ; and those in which the steam and air are not mixed, but expanded separately in separate turbines. So far, in the consideration of mixed-fluid turbines only those employing a working fluid composed of steam and air have been noticed. Proposals, however, have from time to time been made to combine gas turbines with units (regenerative or otherwise) using condensible fluids of low boiling point, such as sulphur dioxide or ammonia. Such schemes are of somewhat a chimerical nature, but they cannot be passed over without mention. They will be considered subsequently.

THE STEAM-AND-GAS TURBINE: MIXED WORKING FLUID.

Turbines of this type, precisely as those of the single-fluid type, may have their range of expansion of the working fluid wholly above the atmospheric pressure, or wholly below that pressure, or partly above and partly below. From the thermal point of view it is irrelevant where the expansion occurs. The Armengaud and Lemale turbine (which approximated to the steam-and-air turbine, if it could not, indeed, actually be regarded as such) expanded down from 6 or 7 atmospheres to atmospheric pressure. Why this expansion was not carried out below the atmosphere instead of above it, the pump being used as an exhauster, is hard to understand. The only objection to using a sub-atmospheric expansion range is the fact that the range is limited to some eight or ten expansions; but with the Armengaud and Lemale machine the pump used was not built for more than seven expansions at the outside. The advantages of working with an expansion range entirely below the atmosphere are very great. First, the difficulties attendant upon continuous combustion under pressure are eliminated. Secondly, the weight of the machine as a whole is considerably reduced, as the maximum pressure difference in the turbine under sub-atmospheric expansion cannot exceed 14 lbs. per square inch. Thirdly, the thermodynamic efficiency of the pump is raised, as the loss due to air friction is directly proportional to the density of the medium through which the pump rotor revolves; as by working below the atmosphere the mean density of the medium would be reduced to one-seventh of that used working above the atmosphere, the friction losses would be one-seventh in the former case to what they would be in the latter. If more than ten expansions are required in the turbine, recourse must be made to super-atmospheric expansion; if not, it is obvious that sub-atmospheric expansion is eminently to be preferred.

The Armengaud-Lemale turbine working above the atmosphere will be considered later, when a review of the experimental work done in recent years on the gas turbine will be examined. It is not very clear whether, in the Armengaud-Lemale experiments, the addition of steam to the working fluid was made in such quantity as to effect the efficiency of the machine and to bring about that increase in overall efficiency which the maximum steam dilution would warrant. It is doubtful, therefore, whether this turbine can be considered as a mixed-fluid turbine proper. It would seem that the experimenters only used steam injection as a means of reducing temperature and not with any thought of a saving in fuel consumption. The constructional details of this turbine and an account of its performance are included in Chapter XI.

THEORIES OF MIXED-FLUID EXPANSION.

The question of expansion of more than one fluid through a nozzle, and the subsequent change of potential energy into kinetic energy, presents considerable difficulty to any satisfactory analysis. The chief difficulty arises from the fact that owing to the difference in value of the specific heats for steam and air, there is considerable difference in velocity between the two fluids if each undergoes a transformation of heat energy to kinetic energy over the same expansion ratio. If the steam and the air were expanded in separate nozzles and upon separate wheels, this difference in velocity would be of no account, as each wheel could run at the velocity of its own working fluid, and there would be no loss of kinetic energy. With the two gases, however, intimately mixed together, and expanded through the same turbine, two theories of the behaviour of the gases present themselves.

I.—The Theory of Common Specific Heat.

The mixture of the two gases may behave as though it were a single gas with a specific heat of such a value as to represent the proportional mean of the specific heats of the two gases of which the mixture is composed. The argument then follows, that the velocity produced would be such as to give in kinetic energy the same value as that given by the sum of the kinetic energies of the two fluids expanded separately, with the corollary that after expansion the internal energy of the mixed fluids would be the same as the internal energy of the two fluids expanded separately.

The theory resolves itself into this: If the two fluids be in the proportion (by weight) of m_1 to m_2, and the velocities of the fluids expanded separately be respectively v_1 and v_2, and the common velocity of the fluid when mixed be v, then—

$$m_1v_1^2 + m_1v_2^2 = (m_1 + m_2)\,v^2.$$

In this case there is no loss of kinetic energy.

There is, however, no evidence to support the theory that the mixed gases behave in this manner, and any such behaviour would tend to conflict with Dalton's Law of partial pressures.

II.—The Theory of Molecular Momentum.

Suppose the two gases of the mixture to behave independently of one another, in accordance with Dalton's Law. Then

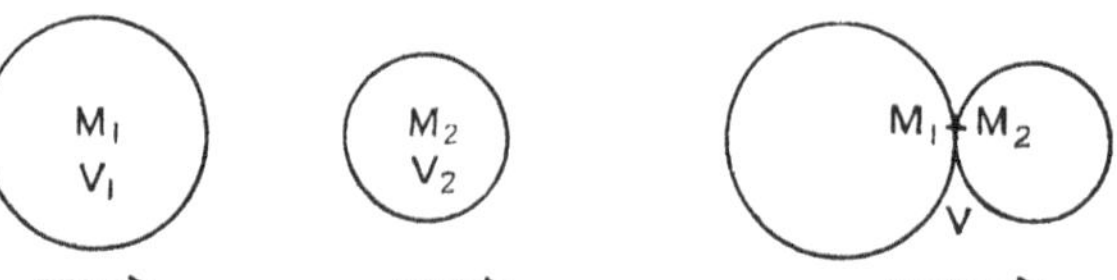

Fig. 25.—Diagram Illustrating Theory of Molecular Momentum.

with the change from heat energy to kinetic energy, each separate molecule starts off with the velocity due to the heat drop of the gas of which it is a constituent part. Let the molecules of one gas be travelling with a velocity V_1 and the molecules of the other gas with a velocity V_2. Let the mass of the molecules be M_1 and M_2 respectively. The gases are intimately mixed. The molecules of one gas are travelling at a different velocity to those of the other gas. The faster moving molecules, therefore, collide with those that are moving at a lesser velocity. Consider two molecules only (Fig. 25). Let the velocity V_1 be greater than the velocity V_2, and let the molecule of mass M_1 with the velocity V_1 approach molecule M_2, which is travelling with the velocity V_2, both molecules travelling in the same direction. After impact has occurred the two molecules of combined mass $(M_1 + M_2)$ may be considered to travel together with a velocity equal to V. Then we have—

$$M_1V_1 + M_2V_2 = (M_1 + M_2) V.$$

It is clear that the values for M_1 and M_2 must be taken in the proportion by weight that the two fluids bear to each other. It will be seen here that there may be a considerable loss of kinetic energy.

The total kinetic energy of the two gases before mixing would be—

$$\tfrac{1}{2} M_1V_1^2 + \tfrac{1}{2} M_2V_2^2$$

the total energy of the gases after mixing has occurred would be—

$$\tfrac{1}{2} (M_1 + M_2) V^2$$

but—

$$V = \frac{M_1V_1 + M_2V_2}{M_1 + M_2}$$

therefore total energy of mixed fluid is—

$$\frac{\tfrac{1}{2} (M_1V_1 + M_2V_2)^2}{M_1 + M_2}.$$

If the ratio of V_1 to V_2 is large the loss in kinetic energy becomes great, but if the value of V_1 approaches that of V_2, then the loss is almost negligible.

The amount of this loss of kinetic energy may be seen by adopting arbitrary values for mass and velocity. Let (for example) $M_1 = M_2$, and $V_1 = 2V_2$.

Then energy before impact of molecules equals—

$$E_1 = \tfrac{1}{2} M_2V_2^2 + \tfrac{1}{2} M_2V_2^2 . 4 = 5/2\, M_2V_2^2.$$

The energy after impact—

$$E_2 = M_2 . 9/4\, V_2^2.$$

Therefore, we have—

$$E_1 : E_2 :: 10 : 9$$

and there is a loss of 10 per cent. in kinetic energy owing to the admixture of the two gases. The loss is, of course, absorbed in the form of heat energy, and the gas is superheated; that is to say, the temperature after expansion is actually higher than the value calculated from the expansion ratio. Some of this energy is recoverable. The case is similar to that of friction loss in the De Laval nozzle. In the "PV" diagram the expansion line is raised out of the true adiabatic towards the isothermal, or in the temperature-entropy diagram the expansion line is shifted out of the vertical. In Fig. 26 the curves XA and YB represent the absorption and rejection of heat at constant pressure on the entropy-temperature diagram. The vertical AK represents the adiabatic expansion. If, owing to frictional losses, the gas is superheated, the expansion curve takes some such form as shown in AB. The extra work done is shown by the area AKB, and the additional heat put in by the area ABCD. It will be seen that the thermal efficiency of the system has been lowered owing to the fact that the area added to the diagram is the least efficient portion of the whole.

With steam and air composing the working fluid, there is an added complication owing to the fact of the difference in temperature of the two fluids after expansion. A heat balance has to be struck between the two gases, a constant mean temperature being arrived at; this is independent of any temperature changes arising out of internal or external friction. Let the temperature of the steam after expansion be θ_1, and that of the air θ_2, let the proportion by weight of air to steam be $1 : y$, and let the specific heats at constant pressure of the two gases be C_a for the air and C_s for the steam. Then the final temperature of the mixed fluid after expansion will be—

$$T = \frac{C_a \cdot \theta_2 + y \cdot C_s \cdot \theta_1}{1 + y}.$$

In this heat exchange there is no energy loss.

The heat changes taking place due to temperature difference and velocity difference with mixed fluids of varying heat capacities is shown in the entropy-temperature diagrams (Figs. 26 and 27).

Fig. 26 is for the air; Fig. 27 for the steam. The steam, after expansion, being at a higher temperature than the air, parts with some of its heat to the air. The quantity of heat thus added to the air is shown by the area ABCD (Fig. 26). The air cycle is thus rendered less efficient. The quantity of heat parted with by the steam is represented by the area PMRT (Fig. 27); the steam cycle thus becomes more efficient.

The loss in thermal efficiency in the one case balances the gain in efficiency in the other case, and there is no change in energy.

The effect of the velocity difference between the two fluids and the subsequent energy loss owing to the concussion of the molecules with the corresponding rise in temperature is shown in the case of the air by the curve AE (Fig. 26); the extra work done is represented by the area AEB, and this work is done at the lowest efficiency on the diagram. The thermal

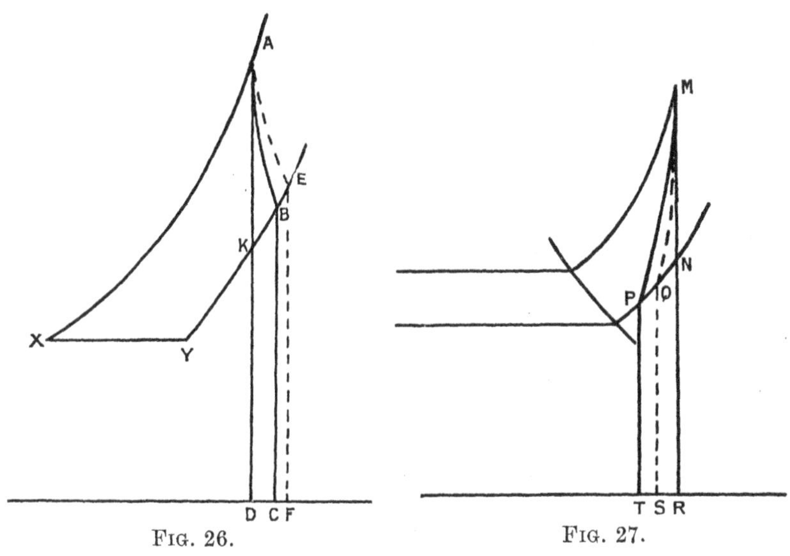

Fig. 26. Fig. 27.

Figs. 26 and 27.—"ΘΦ" Diagrams for Air and Steam, showing Influence of Heat Exchange between two Fluids.

efficiency of the whole is consequently lowered. A similar change takes place with the steam diagram. Owing to the rise in temperature, the curve MP (Fig. 27) is deflected into the position MQ, an efficiency decrease of the same nature being experienced.

In the case of steam and air this loss is practically negligible, as the velocity difference is not great enough to give any large discrepancy between the mean of the squares and the square of the mean value. In an actual case (*vide infra*), with an initial temperature of 1,000° C., and an expansion ratio of 10,

E 2

and an air to steam ratio of 1 : ·45 by weight, the velocity of the air was 3,660 ft./sec., and the velocity of the steam 5,550 ft./sec. This gave a value of work done on separate expansion of 303 T.U., and a value of work done with the two fluids mixed of 299 T.U. The loss due to internal friction (concussion) being about 1 per cent.

This question of " momentum loss " is not of great moment here and it will be neglected in calculating subsequent steam-and-air turbine efficiencies, the loss in question being assumed to be contained in the value taken for the thermodynamic efficiency of the turbine. The matter, however, is of importance in connection with various schemes that have been proposed for injecting high velocity fluids into mediums moving with a lower velocity in the case of explosion turbines (*vide infra*).

THE DAVEY STEAM-AND-AIR TURBINE.

For some years the author has been engaged in considering the question of the practicability of the mixed-fluid turbine as a practical machine. It is with some diffidence that he here prefixes his name to the steam-and-air turbine, but as far as he is aware the subject has not by others been subjected to any very definite investigation. In 1907 the author took out two patents (Nos. 24,085 and 26,580) for gas turbines of this nature. The use of steam as a dilutant is, of course, as old as the gas turbine itself, and there is no claim made as to the novelty of this procedure ; but no one seems to have considered scientifically and in detail the thermodynamic value of the system, the nearest approach to such being a patent of De Ferranti's (No. 13,199 ; 1903) in which the proposal is made to use a " condensible fluid " to reduce the size of the compression plant " because the efficiency of compression in such a plant is comparatively low." The matter, however, is treated in the vaguest manner, the drawings accompanying the patent being merely diagrammatic in nature. Only expansion above the atmosphere seems to have been contemplated and initial pressures of 300 to 400 lbs. are spoken of. There is also in the preamble a rather astonishing disclaimer as to using anything more than a proportion of " gaseous products of combustion amounting at most to only one-third of the working fluid." As the heat value of 1 lb. of combustion mixture (Mond gas) is about 550 T.U., and the latent heat of evaporation of water at 100° C. is 539 T.U., the maximum quantity of water that could be evaporated and raised even to so low a temperature as 100° C. is only ·83 lb., while the proposal was to raise 2 lbs. of steam per lb. of air to a high degree of super-heat. The subsequent patents of de Ferranti (viz., 1409, 1904; 19,912, 1905, and 8886, 1906) deal solely with single-fluid gas turbines, and the earlier system does not seem to have been considered any further.

The practicability of a steam-and-gas turbine as a prime mover competitive with other prime movers rests in three main issues: (1) fuel consumption referred to fuel cost; (2) initial cost; (3) reliability and durability referred to maintenances charges. The fuel consumption depends on three factors: (1) the temperature that the turbine blades can be made to withstand; (2) the peripheral velocity at which the wheel may be run; (3) the thermodynamic efficiencies of the turbine and pump. The two other issues of capital expenditure and maintenance charge rest on the temperatures employed in the turbine; on the pressures used and on the peripheral speed; and on the expansion ratio, referred to the size of the pump. The last two issues work in opposition to the first; that is to say, the higher the initial temperature, the greater the fuel economy, but the less the reliability and durability of the machine; the greater the expansion ratio the greater the fuel economy, but the greater also the first cost of the pump, and the less the thermodynamic efficiency of the same; the greater the peripheral speed the less the reliability, durability and field of application, though the fuel consumed may be less. To strike between these opposing influences the most advantageous balance is the problem that presents itself for analysis to the investigator in gas turbine research.

The first point to be settled is the temperature which may be allowed upon the turbine blades. A maximum practical limit for this temperature may be fixed at 500° C., as in the case of the single-fluid constant-pressure gas turbine previously considered (*vide* Chapter II.). The question will be further noticed in a subsequent chapter when the question of water-cooling is discussed. It may, however, be pointed out in passing that Mr. Neilson (October 21st, 1904 : Institution of Mechanical Engineers) talked of using a temperature of 700° C. on blades non-water-cooled, and of a temperature limit of 2,000° C. on blades that were water-cooled. It is surprising that statements of so wild a nature as this should have been made before any scientific society ; at the least, it is misleading, and can do no good to the advancement of gas turbine design.

The next factor to be considered is that of peripheral speed. With geared turbines very high peripheral velocities are possible, but the insufficiency of the gearing, both in respect to fuel economy and maintenance, is a drawback. For slow-speed turbines, driving alternators direct, a peripheral speed of, say, 850 ft./sec. may be taken as a practical limit of speed. The question of the expansion ratio used in the turbine depends primarily as to whether that expansion is to be wholly sub-atmospheric or not. If the expansion is to take place wholly below the atmosphere, the expansion ratio is perforce limited to a value of about 10. If initial supra-atmospheric pressures are to be permitted, there is no absolute limitation as to the number of expansions used. If the peripheral speed of the pump rotor is to be kept at a reasonably low figure (and it is advisable that this should be done, owing to heavy friction losses at high speeds (*vide* Chapter VII.)), it is not altogether wise to go beyond some 15 expansions or so, except in the high-speed geared type of turbine, where the pump is coupled direct to the main turbine and an increased margin of speed admits of a decreased margin of pump efficiency. With non-geared turbines, using an expansion ratio of 15 or under, the balance of advantages lies with the turbine that works entirely below the atmosphere. The saving in weight alone, and the advantage of not having to make joints to withstand any pressure more than 14 lbs. per square inch, are obvious. The difficulties attendant upon continuous combustion under pressure are also entirely eliminated. The combustion chamber becomes of the simplest construction, the only portions of the machine that require to be airtight being the regenerator and condenser. There is also an advantage in working the rotary air-pump in a less dense medium, as the loss due to skin friction is directly proportional to the density of the medium in which the pump rotor revolves (*vide infra*).

REGENERATION AND THE MAXIMUM PERCENTAGE OF STEAM.

One of the principal difficulties in the single-fluid gas turbine lies in the fact that what regeneration is to be effected has to be effected by means of air to air. This interchange of heat has always been difficult to produce through the necessary medium of the metal wall. With the mixed-fluid turbine, however, the whole of the regeneration may be effected in the raising of additional steam to add to the working fluid, as it is ascertainable that the maximum proportion of steam that it is possible to add to the air is, at the same time, the most efficient proportion to adopt. The temperature difference required for the conduction of heat through metal walls from air to air may be taken to be 200° C., while that required to bring the heat flow between air and water is only about 100° C.; at the same time, a regenerator of very much less bulk is required. It will be seen that here there is a saving on the regeneration alone of 25 T.U. per lb. of combustion mixture in favour of the mixed-fluid system. The water feeding the combustion chamber is, therefore, passed first through the regenerator, where it absorbs heat from the exit gases; the gas has then to be further cooled by means of a condenser, whence it is removed and compressed to the atmospheric pressure by the rotary pump.

To find the maximum quantity of steam that may be added to the products of combustion in the furnace it is necessary to take into account the following factors:—

(1) The heat value of 1 lb. of combustion mixture: let this be denoted by H (for Mond gas H = 550).

(2) The initial temperature in the furnace, θ_1, and the atmospheric temperature, θ_2.

(3) The temperatures at which the working fluid enters and leaves the regenerator, denoted by t_1 and t_2 respectively.

(4) C_{pa}, the specific heat of air at constant pressure, and C_{ps}, the specific heat of steam at constant pressure.

(5) The total heat of 1 lb. of steam at the initial temperature

and pressure prevalent in the furnace, calculated from the temperature of the water from the condenser. Let this value be denoted by m.

The difficulty that arises in calculating this percentage of steam is that every additional quantity of steam raised by the regenerator and added to the furnace raises the mean specific heat of the fluid passing through the regenerator tubes. This process continues until a balance is struck between the heat passing into the furnace from the regenerator and the heat passing into the regenerator from the furnace. The maximum quantity of steam that it is possible to add to 1 lb. of combustion mixture at any given temperature θ_1, without lowering that temperature, is shown by the formula—

$$y\,m = \mathrm{H} - \mathrm{C}_{pa}\,(\theta_1 - \theta_2) + (t_1 - t_2)\,(\mathrm{C}_{pa} + y\,\mathrm{C}_{ps})$$

where y equals the quantity of steam added in lbs. weight.

THE DETERMINATION OF THE INITIAL TEMPERATURE, θ_1.

The temperature on the blades after expansion (assuming all the expansion to take place on one wheel) is 773° A. The expansion ratio is $10 = r$. For any gas expanded adiabatically we have the equation—

$$\theta_1 = \theta_2 \,.\, r^{\frac{\gamma - 1}{\gamma}}$$

for the products of combustion γ may be taken as equal to 1·38. Therefore—

$$\theta_1 = \theta_2 \,.\, 1{\cdot}884 = 1{,}460° \text{ A.}$$

For steam $\gamma = 1{\cdot}3$; therefore—

$$\theta_1 = \theta_2 \,.\, 1{\cdot}7 = 1{,}320° \text{ A.}$$

Taking the mean of these two values the temperature before expansion of the working fluid may be taken as 1,390° A., or 1,117° C.

The initial temperature having been determined, the maximum quantity of steam to be added can be calculated.

Let the exhaust gases leave the regenerator at 100° above the atmospheric temperature. The total heat for 1 lb. of steam for 15 lbs. pressure and 1,117° C. temperature, reckoned from a condenser temperature of 45° C., is 1,103 T.U.

We have, then, in the above formula for weight of steam added—

$$y\ 1{,}103 = 550 - {\cdot}25\,(1{,}390 - 288) + (773 - 388)\,({\cdot}25 - {\cdot}5y).$$

This gives a value—

$$y = {\cdot}4 \text{ lb.}$$

WORK DONE AND VELOCITY PRODUCED.

The work done by the gas on expansion equals—

$$\int_{p_2}^{p_1} v \, . \, dp - \text{in units of work,}$$

or—

$$C_p\,(\theta_1 - \theta_2) \text{ in heat units.}$$

$$= {\cdot}25\,[1{,}460 - 773] = 172 \text{ T.U.}$$

The work done by 1 lb. of steam is given by the Rankine formula—

$$W = (\tau_1 - \tau_2)\left(1 + \frac{L_1}{T_1}\right) + C_p\,(\tau_s - \tau_1) - \tau_2\left(\log_e \frac{\tau_1}{\tau_2} + C_p \log_e \frac{\tau_s}{\tau_1}\right).$$

This gives a value of work done per lb. of steam of 352 T.U. For ·4 lb. of steam (the weight added) this becomes 141 T.U. Therefore the total work done by 1·4 lbs. of working fluid equals—

$$172 + 141 = 313 \text{ T.U.}$$

VELOCITY.

If W is the work done in heat units per mass, m, of the working fluid, then we have the relation—

$$W = \frac{mv^2}{2g \, . \, J},$$

g being the acceleration of gravity and J the mechanical equivalent of heat. Substituting the value 313 for W and 1·4 for m, this gives a value for the velocity equal to 4,490 ft./sec. As has been previously noted, the small decrease in velocity due to the collision of molecules of the two gases has not been taken into account.

The peripheral velocity of the turbine wheel may be taken to be 850 ft./sec. (*vide supra*). Only one wheel is used, a double wheel of the Curtis type. This is sufficient to take out the greater part of the velocity of the impinging gases. Fig. 28 shows a velocity diagram for the double wheel, drawn for the velocities in question, viz., a peripheral velocity of 850 ft./sec., and an initial velocity of the working fluid of 4,490 ft./sec.

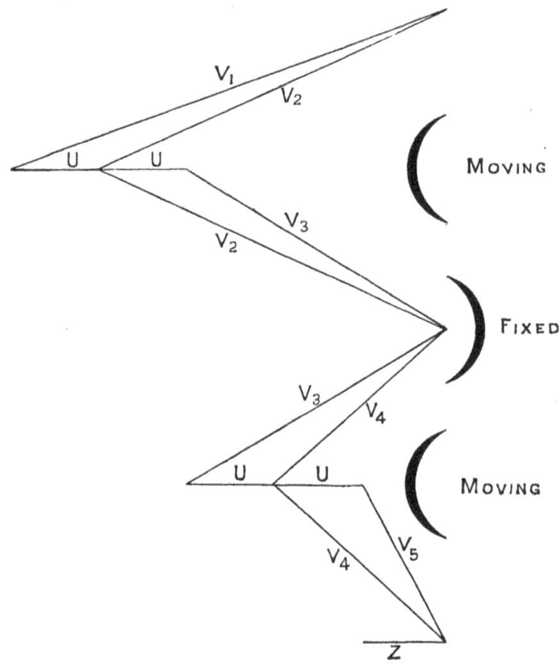

FIG. 28.—Velocity Diagram for Curtis Wheel. $V_1 = 4{,}490$ ft./sec.; $U = 850$ ft./sec.; $V_3 = 2{,}900$ ft./sec.; $V_5 = 1{,}700$ ft./sec.; Z = component in direction of rotation (recoverable) = 800 ft./sec.

The final velocity of the gases leaving the last ring of blades is thus 1,700 ft./sec. The total energy per lb. of working fluid is $k\ 4490^2$ units of work where k is a constant; the energy lost in residual velocity is $k\ 1700^2$ units. This gives a theoretical efficiency for the wheel of 80 per cent.

After passing through the wheel the working fluid passes into the regenerator. This is practically of the same nature as a fire-tube boiler; all the water feeding the furnace is first conducted through this heater. The exhaust gases then pass on to the condenser where the residual heat is abstracted before the fluid passes into the pump. The pump compresses the waste gases from a pressure of 1·47 lbs. per square inch up to the atmospheric pressure. The work done by the pump is given by the equation—

$$W = p_1 v_1 \log_e r.$$

where p_1 and v_1 are the initial pressure and volume and r the compression ratio. This ratio is, of course, the same as the expansion ratio in the turbine. The pump is so cooled between its elements as to insure isothermal compression. With ten compressions the work done on the pump per lb. of products of combustion delivered is 43 T.U.

The thermal efficiency of the cycle equals—

$$\frac{\text{Positive work} - \text{Negative work}}{\text{Heat put in}}$$

this equals—

$$\frac{313 - 43}{550} = 49 \text{ per cent.}$$

The overall efficiency is found by inserting values for the thermodynamic efficiencies of the turbine and pump in the usual manner. Let these values be 70 per cent. for each. Then—

$$\text{Overall efficiency, E} = \frac{313 \times 70 \text{ per cent.} - 43/70 \text{ per cent.}}{550}$$
$$= 28{\cdot}5 \text{ per cent.}$$

If reference is made back to the diagram of efficiencies for the single-fluid turbine on p. 37, it will be seen that the efficiency of the single-fluid turbine for the same initial temperature of the working fluid and the same thermodynamic efficiencies of turbine and pump is 24·5. Thus, by the addition of steam to the products of combustion an actual saving in fuel consumption has been effected amounting to 14 per cent.

It should be borne in mind that the efficiencies plotted in Fig. 18 are based upon an expansion ratio of 20, whereas in the present case the expansion ratio is only ten.

A more detailed comparison between the efficiencies of the single-fluid and the mixed-fluid types of turbines will be considered shortly.

THE PUMP.

The success of the gas turbine as a practical machine rests very largely with the pump. This is practically confined to the rotary variety. To insure the number of elements being kept within reasonable bounds for the ratio of compression required, it is necessary that a high peripheral velocity be achieved by the pump. The formula given by Professor Rateau for his rotary pump (*vide infra*) is—

$$r = 1 + 6.10^{-7} . v^2$$

where r is the compression ratio effected per stage, and v is the peripheral velocity of the rotor in feet per second. With a peripheral velocity as used in the turbine of 850 ft./sec., r becomes 1·434. The compression achieved by a number of elements ascends in geometrical progression, the total compression R produced by a number of elements, n, being equal to r^n, r being the compression ratio for the single element. Thus—

number of elements required,
$n = \log R/\log r$.

With $R = 10$ and $r = 1{\cdot}434$, the value for n equals 6·4 ; this means some seven elements in the pump.

It is not advisable to run the pump on the main turbine shaft, as to insure the peripheral speed of 850 ft./sec., it would be necessary to construct the pump rotors of the same diameter as the turbine wheel, viz. 6·5 feet. This would render the pump of excessive bulk, the frictional surfaces would be greatly increased and the difficulty of controlling the delivery would become very great. It is on all counts more advantageous to drive the pump by a separate high-speed turbine ; in this case the size of the pump can be controlled at will. With a shaft-speed for instance, of 7,000 r.p.m., the diameter of the pump rotor to insure a peripheral velocity of 850 ft./sec. would be only 2 feet 4 inches. With the pump run upon a separate shaft to that of the main turbine there is theoretically no limit to the peripheral speed at which the pump may be run, and it would seem that on this count the number of elements in the pump could be reduced to a minimum at will. There are objections, however, to doing this. First, the losses in the pump due to skin friction are proportional to the cube of the

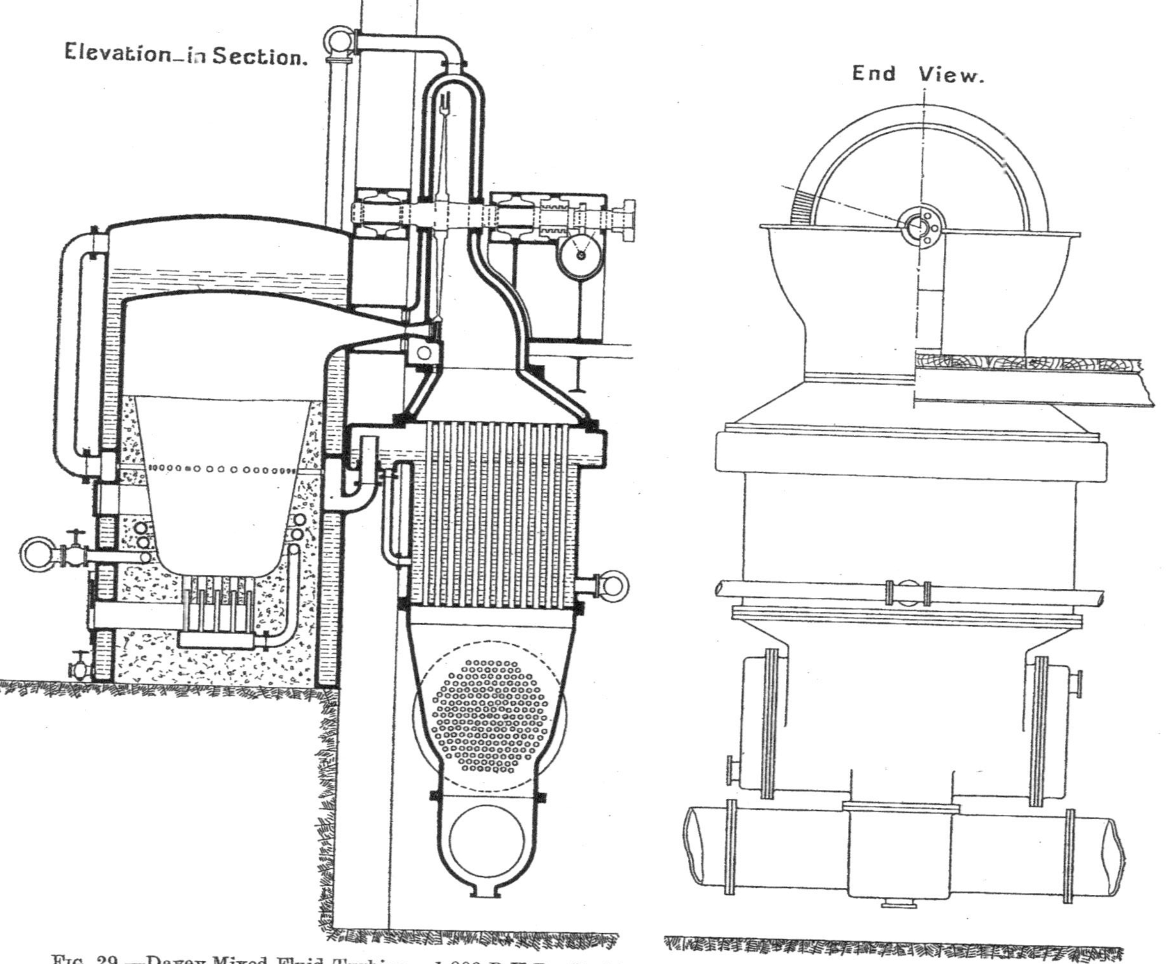

FIG. 29.—Davey Mixed-Fluid Turbine. 1,000 B.H.P. Positive Unit. 2,500 r.p.m. Scale, $\frac{1}{4}$ inch to 1 foot.

peripheral speed, a doubling of the speed producing an increase in these losses to eight times the original value. Secondly, the fewer the elements in the pump for a given ratio of compression, the less perfect is the means afforded for cooling the compressed fluid and the less nearly does the compression approach that of the isothermal ideal. A point is soon reached at which the saving in weight and space produced by lowering the number of elements in the pump is counterbalanced by a decrease in efficiency.

ACTUAL DESIGN OF MIXED-FLUID TURBINE.

In accordance with the example worked out in the foregoing pages, the author has evolved a general design of a mixed-fluid turbine of this type (*i.e.*, sub-atmospheric, with an expansion range of 10, initial temperature, 1,117° C.), which is illustrated in Figs. 29, 30, and 31. The turbine in question is of the single element type, using a double Curtis wheel, with a peripheral velocity of 850 ft./sec. The turbine is designed to run at 2,500 r.p.m.; this gives a rotor diameter of 6·5 feet. Fig. 29 shows a sectional view of the "main" turbine in a plane at right angles to that of rotation. The machine is designed for 1,000 B.H.P., the scale is ¼ inch to 1 foot. Taking the thermodynamic efficiencies of the turbine and the pump to be each 70 per cent., the ratio of the brake horse-power done by the turbine to the brake horse-power required to drive the pump is 220 to 62. If the motive power available be divided into three turbines each of 1,000 B.H.P., to drive the alternators, it will be seen that the fourth turbine unit required to drive the pump (which undertakes the negative work for itself and for the other three units), will be a unit of the same order of power; the actual power required for the unit in question being 1,180 B.H.P. As this is only 18 per cent. larger than the 1,000 B.H.P. units, the general dimensions of the same have been taken as identical with that of the positive power units; more especially as a wide margin has been left in calculating the dimensions for these latter.

THE FURNACE.

It will be seen from Fig. 29 that the furnace in which the fuel is burnt and the working fluid formed by the addition of the steam, is entirely jacketed with water. The water that is fed to this jacket is first led through the regenerator where waste heat is taken up from the exhaust of the turbine. All the heat that is conducted through the walls of the combustion chamber to the water surrounding it simply goes to raise the steam with which the products of combustion are diluted, and by which the temperature of the working fluid is reduced to the limit of practical working in the turbine, this temperature being, as stated above, 1,117° C. It will thus be seen that there is little or no radiation losses, the difference in temperature between the inside of the outer wall of the furnace and the atmospheric temperature being only 85° C., as compared to 1,102° C. were there to be no steam dilution, and consequent water jacket surrounding the combustion chamber.

The furnace is brought straight up to the turbine wheel which enables the nozzle to be water-jacketed, together with the turbine casing itself, which jackets communicate with the main jacket of the combustion chamber. The regenerator, being in the form of a fire-tube boiler, is also almost entirely surrounded with water. In this manner all radiation losses, from whatever part of the working fluid path, are reduced to a minim.ım.

The lower part of the furnace is lined internally with fire brick, and in this portion the gas and air is mixed and burned ; steam is admitted higher up, by a ring of holes in the side of the furnace, the steam being conducted from the space above the water in the furnace jacket.

The sectional area of the interior of the furnace is 24 square feet, which gives a velocity of the working fluid through the furnace of 10 feet a second. The burners are of the Bunsen type and consist of a "nest" of eighteen 3-inch burners in concentric rings, with a velocity through the burners of 30 ft./sec. The pipe conveying the fuel to the furnace is shown with a coil embedded in the fire-clay lining of the furnace, the arrangement being designed for use with oil, preliminary evaporation being secured by the coil. The air admission to the furnace is regulated by a damper in the usual manner. The actual weight of working fluid passing per second for 1,000 B.H.P. under stated conditions is 3·56 lbs. ; this gives a value in volume passed per second of 2,390 cubic feet. The nozzle area required at a velocity of 4,490 ft./sec., would thus

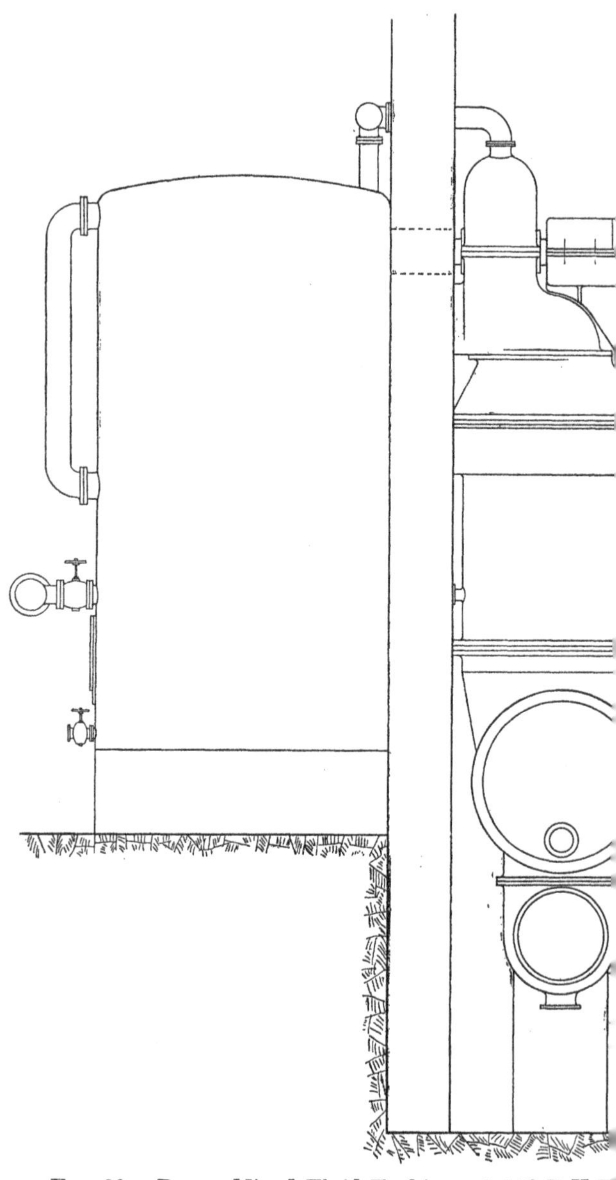

FIG. 30.—Davey Mixed-Fluid Turbine. 1,180 B.H.P

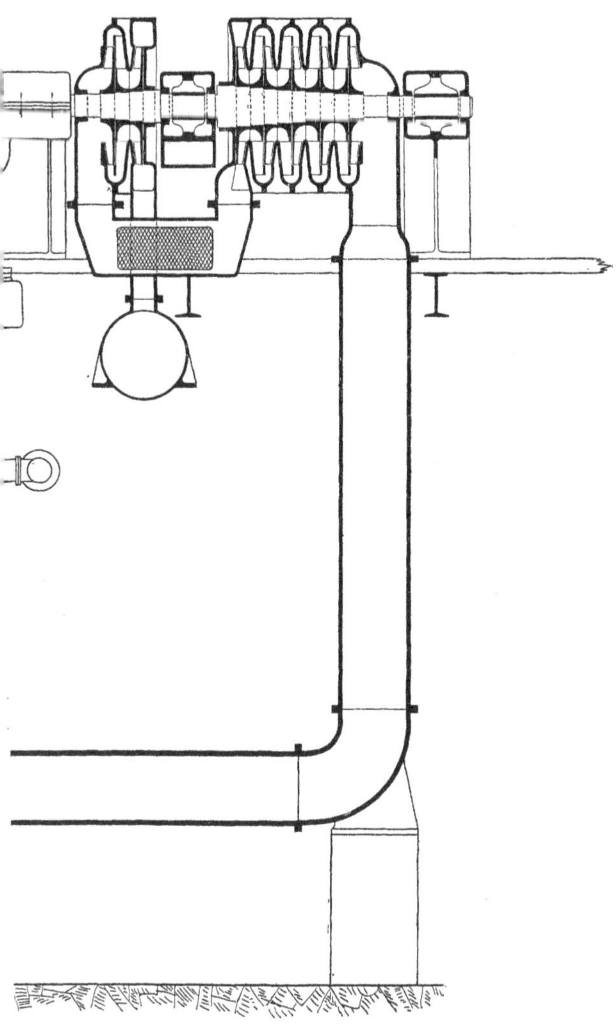

ative Unit. 6,000 r.p.m. Scale, ¼ inch to 1 foot.

be 77 square inches. The inclination of the guide blades to the plane of rotation of the wheel is taken at 20°, and the length of the first ring of blades is 4 inches. This means that the nozzle would occupy 56 inches of the circumference of the turbine wheel, and the turbine blades would only be subject to the effluent gases for ·23 of each revolution.

THE REGENERATOR.

Arguing from the analogy afforded by tests that have been carried out in respect to the relative evaporating power of various portions of a locomotive boiler (Pye Williams, Donkin, etc.), that the heat transmitted is about 2,000 T.U. per square foot of tube surface per hour, the tube surface required for the regenerator is 800 square feet, the exit temperature being taken at 115° C. Taking it that the T.U.'s passing in the condenser per square foot per second equals 1·5, the surface of the condenser tubes required is 433 square feet.

An end view of the turbine is shown projected on to the section. This end view is not shown in section, but the cover of the turbine casing is shown removed. The exhaust gases from the condenser pass into the exhaust pipe that runs along under all the power units in the building, every fourth unit being provided with a pump whereby the pressure difference is maintained.

Fig. 30 shows a turbine unit coupled up to an exhausting pump. The scale is the same as in the preceding figure. Only the pump itself and the pipe connections to the condenser are shown in section. The heights of the three floors—the turbine floor, the furnace floor, and the condenser floor—are drawn at the same height as in the preceding figure. The pump is designed to run at 6,000 r.p.m., with a peripheral speed of 850 ft./sec. This gives a diameter of the rotor of 2 feet 8 inches for both turbine and pump. It will be seen that the size of the turbine is thus considerably reduced, and the furnace and regenerator are both raised above the level of those in the positive power units. It has been seen (p. 60) that the value of n in the pump equation—

$$R = r^n$$

is 6·4; R being the total expansion ratio, viz., 10, and r

the ratio of compression for a single element, being, at 850 ft./sec. peripheral speed, equal to 1·434. This means that there are to be some seven elements in the pump.

BALANCING.

The pump has too many elements to admit of perfect balancing by the method of duplicating the rotors and delivering the gas through the pump in opposite directions. The nearest approach to perfect balance is to arrange the elements of the pump in such a manner that the first five wheels deliver the air in one direction and the last two in the opposite direction. The pump is divided in this way in Fig. 30; the thrust left over being opposed to the thrust set up by the turbine owing to the axial component of the velocity of the working fluid. The difference in pressure on the rotor discs between the two divisions of the pump is ·43 lbs. per square inch of disc surface; the average disc surface is 740 square inches. This gives a thrust on the shaft of 320 lbs. This is not excessive and, in addition, is partly counterbalanced by the thrust from the turbine.

COOLING.

The pump is of the Rateau type, and is water-jacketed; the water jacket reaching into the centre of the pump almost up to the inner edge of the rotor blades; effective cooling is thus insured. This is further secured by the interposition of a cooler between the two portions of the pump through which the gas is made to pass. A condition of compression approaching the isothermal is thus brought about.

GENERAL ARRANGEMENT OF PLANT.

In Fig. 31 is shown the general arrangement of the gas turbine plant. There are six "positive units" each of 1,000 B.H.P. The necessary "negative" work for this plant is performed by two pump units each of 1,180 B.H.P. They are placed at each end of the building and are so arranged as to act as exhausters for three turbines only; in this manner one-half of the plant can, at any time, be shut down, without effecting the operation of the other half. The turbine units are

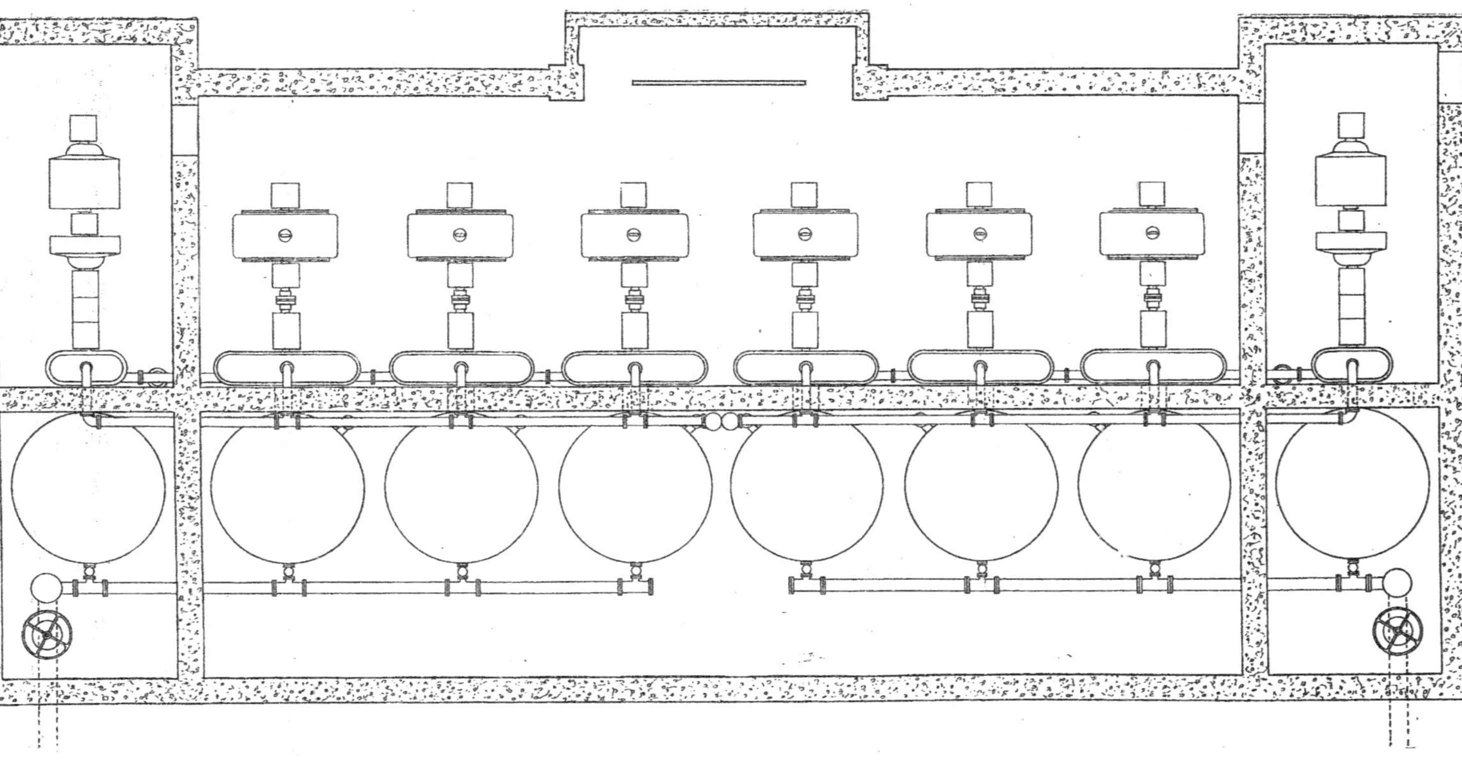

FIG. 31.—Davey Mixed-Fluid Turbine. Plan of General Arrangement of Plant. 6,000 B.H.P. Scale, $\frac{1}{8}$ inch to 1 foot.

shown driving dynamos; the whole plant represents a *net* output of 6,000 B.H.P. In this plan the circulation of the jacket water is shown. The positive turbine units exhaust into a large pipe that runs the whole length of the building; the pumps exhaust from this, there being no connection between the three right-hand turbines and the three left-hand turbines. The regenerators, condensers, and piping appertaining thereto lie beneath the engine-room floor, and are not shown in the plan.

Figs. 29, 30 and 31 are all drawn for turbines burning oil fuel. With gaseous fuel some modifications are necessary. As regards the type of steam and oil turbine shown in Figs. 29, 30 and 31, this means little more than an increase in the size of the piping and valves supplying and admitting fuel to the furnaces. In regard to the question that crops up with the use of gaseous fuel in steam and gas turbines, in connection with the formation of sulphuric acid in the regenerator, the reader is referred to a subsequent chapter (*vide* p. 163) where the matter is treated more fully. It may, however, be pointed out here that, with steam and air turbines in which there is initial compression of the fuel before combustion, the oil turbine has one advantage over the gas turbine. Power gas requires for its complete combustion somewhere about its own volume of air. As it is not permissible to mix the air and the gas before combustion, it is necessary to provide *two* rotary pumps, one for the air and one for the gas. The cost of rotary pumps is not in direct proportion to the output of the pump, a pump of 500 H.P. being almost as costly as one of 1,000 B.H.P. The case presents itself as follows: For the oil-burning turbine a rotary pump of (say) 1,000 B.H.P. will be needed together with a small oil piston pump of trifling size and cost. For the gas-burning turbine, two pumps of 500 B.H.P. each will be necessitated of almost equal size and cost to the 1,000 B.H.P. pump. One advantage that the sub-atmospheric type has, therefore, is its adaptability to either gas or oil fuel with trifling alterations.

GOVERNING.

The governing of the steam-and-air turbine working below the atmosphere may be effected by two methods: (1) by the lowering of the temperature of the working fluid before expansion through the turbine; (2) by "breaking the vacuum" in the condenser, *i.e.*, by lowering the expansion ratio (*vide* Chapter VII.). Either method affects the efficiency of the machine while the controlling of the governor is taking effect. In this respect it may be noted that the explosion turbine (*vide infra*) has the advantage of the continuous flow type, as the number of explosions per given unit of time may be varied, and the power of the machine thus altered, without altering

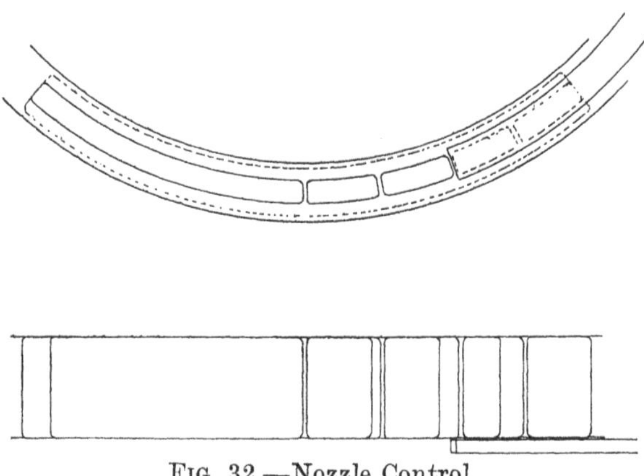

FIG. 32.—Nozzle Control.

the limits of temperature or pressure upon which the efficiency of the turbine depends.

LOAD VARIATION.

The practicability of the variation of the load in the case of the steam-and-air turbine depends upon two factors: firstly, the limits of nozzle control; secondly, the variation in the pump efficiency with variable output.

In Fig. 32 is shown a system of nozzle control, from full load to half load, in one-eighth-load steps, by a sliding shutter cutting off the nozzles. The dimensions of the plant apply, of course, for loads less than full load, the velocity of the fluid through the various channels simply being reduced.

If reference is made to the graphs on p. 174, it will be seen that for a Rateau pump the variation of efficiency over a load of 1 to ·5 is about 4 per cent. In the example here considered the pump efficiency at full load was taken at 70 per cent. At half load the efficiency would then be 66 per cent., and the negative work done on the pump would rise from 62 T.U. per lb. of air exhausted to 65 T.U. This would reduce the overall efficiency from 28·5 to 28·2, resulting in a nett rise in fuel consumed per unit of work done on the shaft of about 1 per cent. It will thus be seen that the loss in efficiency over varying loads is really not excessive.

MIXED FLUID TURBINES WORKING WITH SUPER-ATMOSPHERIC EXPANSION.

As has been pointed out above, a mixed-fluid turbine working below the atmosphere is limited to an expansion ratio of 10. If it be proposed to use initial compression of the working fluid, it would be theoretically possible, by substituting two rotary pumps for one and working those two pumps in series—the one as an initial compression pump, the other as a final exhausting pump—to achieve an expansion ratio of 100 by the duplication of the pumping machinery required for ten expansions. A duplication of the wheels of the turbine would also serve to take up work evolved by the 100 expansions, the initial temperature remaining the same as before. Taking approximate figures this would produce an additional temperature drop of 330° C. in addition to the first temperature drop of 620° C. produced by the initial ten expansions. This represents, roughly, an increase in the gross work of 50 per cent. As, however, the pump is compressing isothermally while the turbine is expanding adiabatically, the ratio of the negative to the positive work for the first range of ten expansions will be less than the ratio in the second range of ten expansions. With the ten expansion range ($\theta_1 = 1{,}117°$ C.) the positive work, at thermodynamic efficiency for turbine of 70 per cent., equals 220 T.U. per lb. of combustion mixture, and the negative work ($e_p = 70$ per cent. for pump) is 62 T.U. We may take it roughly that the positive work is increased to 300 T.U. and the negative work to 100 T.U. The heat put in equals 550 T.U. The overall efficiency, therefore, is—

$$\frac{300 - 100}{550}$$
$$= 36 \text{ per cent.}$$

This is only an approximate deduction. It must be remembered that the final theoretical temperature of the fluid after the 100 expansions would be 167° C., a temperature too low to afford regenerative heat. The quantity of steam added to the working fluid would therefore be much reduced, a reduction, in point of fact, of ·4 lb. of steam per lb. of products of combustion to ·25 lb. This increases the ratio of the negative to the positive work, an increase that has not been taken into account in the above approximation (*vide infra*).

A compression ratio, however, of 100 is not attainable, or, at least, has so far not been attained with rotary pumps. Though in theory the coupling together in series of two pump units giving x compressions produces a total compression of x^2, in practice there is a considerable falling off, when the ratio of compression is largely increased. For more detailed information as to this and other matter relating to the rotary pump the reader is referred to Chapter VII.

An expansion ratio, however, of, say, 30 may be taken as a conceivably possible figure. It may be of interest to work out a case of this kind, assuming θ_1 to be, as before, 1,117° C. or 1,390° A.

Let the initial pressure be 60 lbs. per square inch and the terminal pressure be 2 lbs. The temperature, θ_2, of the air after expansion is given by the formula—

$$\theta_2 = \frac{1{,}390}{30^{\frac{\gamma-1}{\gamma}}}$$

γ being equal to 1·38 for products of combustion; for steam γ becomes 1·3.

This gives a temperature for the air after expansion of 272° C., and for the steam 362° C. Assuming the ratio of the air to the steam to be 3 to 1, we get a mean final temperature of 340° C.; that is to say, the entry temperature to the regenerator is 340° C.; let the exit temperature from same be (as before) 115° C. From the formula given on p. 56, the maximum quantity of steam that can be added to the products of combustion is found to be ·32 lbs.

The work done by 1 lb. of steam expanding from 60 lbs. down to 2 lbs. with an initial temperature of 1,117° C. is, from the Rankine formula, 417 T.U. The work done by 1 lb. of products of combustion over the same range of expansion and with the same initial temperature is 211 T.U. The total work, therefore, done by 1·32 lbs. of working fluid, composed of a mixture of 1 lb. of air and ·32 lb. of steam, is 345 T.U. The negative work done by the pump per lb. of air compressed is—

$$\omega = \frac{p_0 v_0}{J} \log_e r \text{ T.U.}$$

$$(r = 30,)$$

therefore work done by pump equals 64 T.U.

Therefore thermal efficiency—

$$\eta = \frac{344 - 64}{550}$$

$$= 51 \text{ per cent.}$$

It will be seen that this is only a slightly higher thermal efficiency than that obtained with ten expansions, the thermal efficiency in that case being 49 per cent. Taking the thermodynamic efficiencies of the turbine and pump to be each 70 per cent., the overall efficiency of the machine becomes 27·2 per cent. The overall efficiency in the case of the turbine with ten expansions was 28·5. It will be seen that this change in relative advantage between the thermal efficiency and the overall efficiency of the two cases is owing to the fact of the larger proportion that the negative work bears to the heat put into the system in the latter case (thirty expansions) than in the former (ten expansions).

The approximation of the thermal efficiencies in the two cases is due to the fact that, owing to the lowering of the temperature of the gases from the exhaust of the turbine, the quantity of heat available in the regenerator is considerably reduced. It must be borne in mind that all the heat liberated in the regenerator goes to the raising of additional steam to the products of combustion. The quantity of steam added is reduced from ·4 lb. in the former case to ·32 lb. in the latter. The work done by the quantity of steam added in the former case is—

$$352 \times \cdot 4 = 141 \text{ T.U.}$$

and that added by the steam in the latter case—

$$417 \times \cdot 32 = 133 \text{ T.U.}$$

There is thus a nett loss in the latter case of work done by the steam of 8 T.U.

In the former case the work done by the air is 172 T.U. In the latter case 211 T.U. The nett gain in positive work is therefore 21 T.U. on the side of the thirty expansion case. The negative work, however, is, with ten expansions, 43 T.U., and with thirty expansions, 64 T.U. There is a nett gain here on the side of the ten expansion case of 21 T.U. Balancing these two gains against one another, any advantage that the one case has over the other is eliminated. As has been seen above, when the overall efficiency is taken into account, the advantage lies on the side of the fewer expansions.

It will thus be seen that the increase in the expansion ratio, *provided the initial temperature of the working fluid be kept constant,* produces no corresponding increase in the overall economy of the system, and that with ten expansions in the turbine a higher overall efficiency is produced than with thirty expansions. It is obvious, therefore, that the turbine working *wholly* below the atmosphere possesses every advantage over any other type having additional initial compression ; as the only result produced by such is increased range of expansion, and this eventually *lowers,* instead of increases, the overall efficiency. The practical advantages that lie with the sub-atmospheric type have already been noted ; it may also be borne in mind that not only is the pump less bulky, owing to the fewer number of compressions that it has to perform, but also its thermodynamic efficiency is higher without initial super-atmospheric compression than with it, as the mean density is less in the former case than in the latter, and the losses due to disc friction are directly proportional to the mean density of the medium in which the discs rotate.

The author has felt that the statement just made, viz., that, within definite limits, the overall efficiency of the steam-and-air turbine changes in an *inverse* proportion to the expansion ratio in the turbine, is so (apparently) in direct antithesis to the fundamental laws of thermodynamics that some further notice of the phenomenon is necessary.

It has been already shown that the addition of steam to the products of combustion produces an increased overall efficiency within certain definite limits of maximum temperature used.

What those limits are will be shown later. It is now to be shown that, with a certain fixed value for the initial temperature of the cycle and fixed values for the thermodynamic efficiencies of turbine and pump, an increase in the expansion ratio is actually accompanied by a decrease in *nett* work available in the form of brake horse power on the turbine shaft, that is to say, by a decrease in the overall efficiency of the system, as the heat put into the furnace remains constant.

One of the terminal pressures has to be fixed; the lower terminal pressure has been chosen for the fixed value; this remains constant at 2 lbs. per square inch absolute. Values for the upper limit of pressure have been taken at 14·7, 20, 40, 60, 80 and 100 lbs. per square inch, giving expansion ratios of 7·3, 10, 20, 30, 40 and 50 respectively. The initial temperature of the cycle is fixed at 1,390° A.

The problem is treated analytically in the table on p. 72.

The temperature of the working fluid is first found for the various expansion ratios after the expansion has taken place. This temperature will have a different value for the air and steam; these two values are found separately, in the ordinary way, taking $\gamma = 1{\cdot}38$ for the products of combustion, and $\gamma = 1{\cdot}3$ for the steam; the mean temperature is then taken. This gives the temperature of the fluid upon leaving the turbine and of its entry to the regenerator. These values will be found in the second, third and fourth columns in Table I., p. 72. The fifth column gives the value of m, which is the total heat of 1 lb. of steam reckoned from the condenser temperature for the various initial pressures.

The temperature at which the exhaust gases leave the regenerator is fixed at 115° C. This, together with the values of m and θ_m, gives the value of y, the maximum quantity of steam that may be added to the products of combustion without lowering the fixed initial temperature in the furnace. It will be noticed that this value decreases with the increased range of expansion, owing to the lowering of the terminal temperature, which reduces the heat absorbed in the regenerator and consequently the quantity of steam that the regenerator adds to the furnace. In Column VIII. is given the work done (in heat units) by the expansion of 1 lb. of steam over the given expansion range. Column IX. gives the multiple of W_s and y,

TABLE I.

Number of Expansions.	θ_{2_a}	θ_{2_s}	θ_{2_m}	m	Regenerator $(t_1 - t_2)$ $t_1 = \theta_{2m}$ $t_2 = 388°$ A.	y	W_s	$W_s . y$	W_a $\frac{\theta_1 - \theta_2}{4}$	W_n	Net work. (W) $(W_s . y \times W_a - W_n)$	Net Work. (P . 70 per cent.) (T . 70 per cent.) $= \cdot 7 W - \frac{W_n}{\cdot 7}$	E
I.	II.	III.	IV.	V.	VI.	VII.	VIII.	IX.	X.	XI.	XII.	XIII.	XIV.
7·3 14·7—2	810	880	833	1,095	445	·44	369	162	145	37	270	161	29·2
10 20—2	738	820	760	1,094	372	·40	380	152	163	43	272	160	29·1
20 40—2	610	700	635	1,091	237	·34	401	136	195	56	275	152	27·6
30 60—2	545	635	567	1,087	179	·32	415	132	211	64	279	149	27·1
40 80—2	505	593	525	1,086	137	·30	422	127	221	69	279	145	26·4
50 100—2	473	565	493	1,085	105	·29	424	123	229	73	279	143	26·0

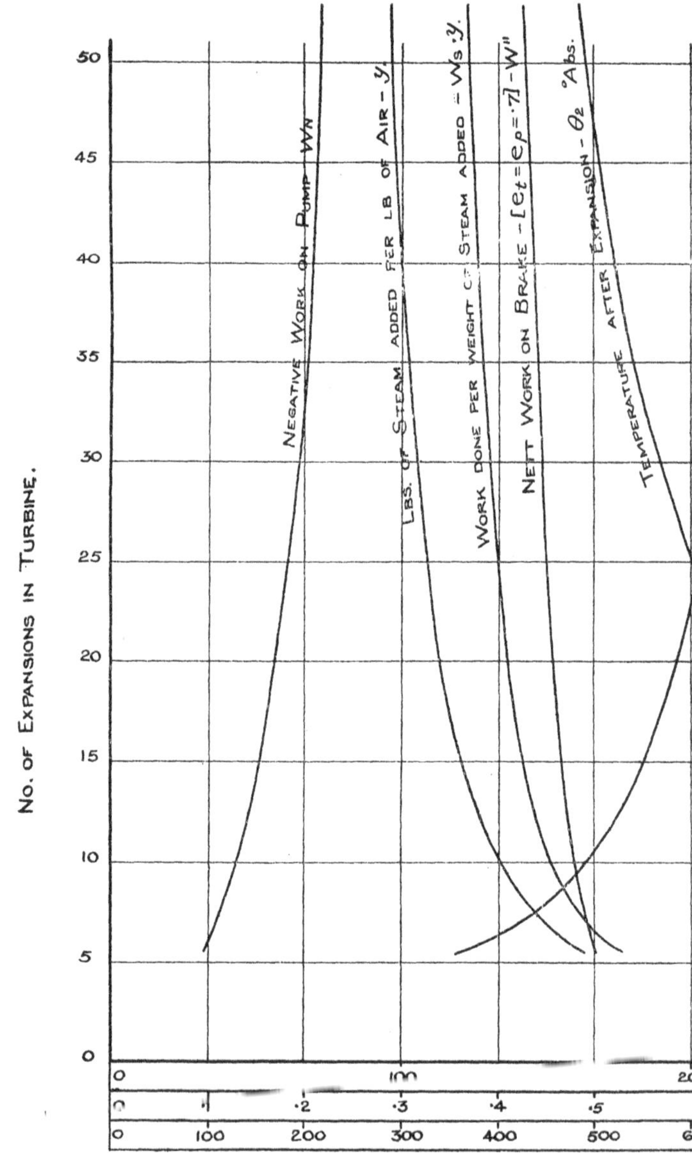

Fig. 33.—Diagram showing Change in Work

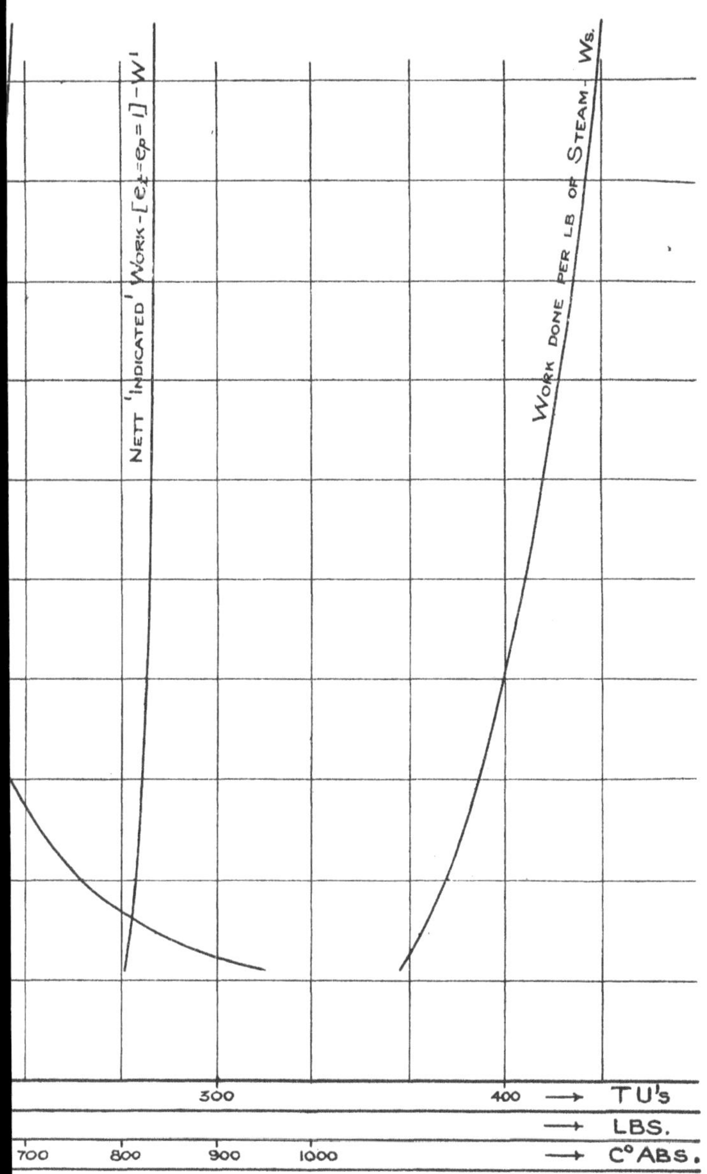

er lb. of Fluid with Increase in Expansion Ratio.

or the work done by the steam in the system per lb. of products of combustion. It will be noticed that this value shows a decrease with the increase in expansion ratio, owing to the diminishing value of y.

Column X. gives the value of the work done per lb. of air in expansion, Column XI. the work done on the air in compression over the same range. It will be noticed that both W_a and W_n increase with the expansion ratio, but that owing to the fact that the turbine expands adiabatically while the pump compresses isothermally the increase of W_n is proportionately greater than the increase of W_a. The nett work, assuming no losses in either turbine or pump, is given in column XII. ; there is an increase with the expansion ratio, but this increase is very slight, amounting to only 3·3 per cent. between the ratio limits of 7·3 and 50. In Column XIII. is given the nett work available on the turbine shaft in thermal units per lb. of products of combustion, taking into account the losses in the turbine and pump and assuming these losses to amount in each case to 30 per cent. It will be seen that the nett work actually available *decreases* with the increase in expansion ratio. This is owing to the larger proportion of negative work with the increased expansion. As the heat put into the furnace remains constant ($\theta_1 = 1{,}390°$ A = constant) the overall efficiency decreases with the increase in expansion ratio. These values are given in Column XIV.

The variation of these factors with the change in expansion ratio is further shown by the curves in Fig. 33. The diagram is a composite one, representing the change in terminal temperature, θ_2; quantity of steam added, y; work done per lb. of steam, W_s; work done per y lbs. of steam, $y.W_s$; work done per lb. of air, W_a; negative work done per lb. of air, W_n; nett "indicated" work, W'; and nett available work taking into account turbine and pump losses, W'', for varying values of the expansion ratio from 7·3 to 50.

It is interesting, however, to see the change in overall efficiency for the mixed-fluid turbine when the expansion ratio is kept constant and the initial temperature is altered. This change is shown in the diagram (Fig. 34) by the full line curves. The top curve represents the change in thermal efficiency with the rise in initial temperature. The three lower curves represent the change in overall efficiency for the three conditions of thermodynamic efficiency: (1) turbine 75 per cent., pump

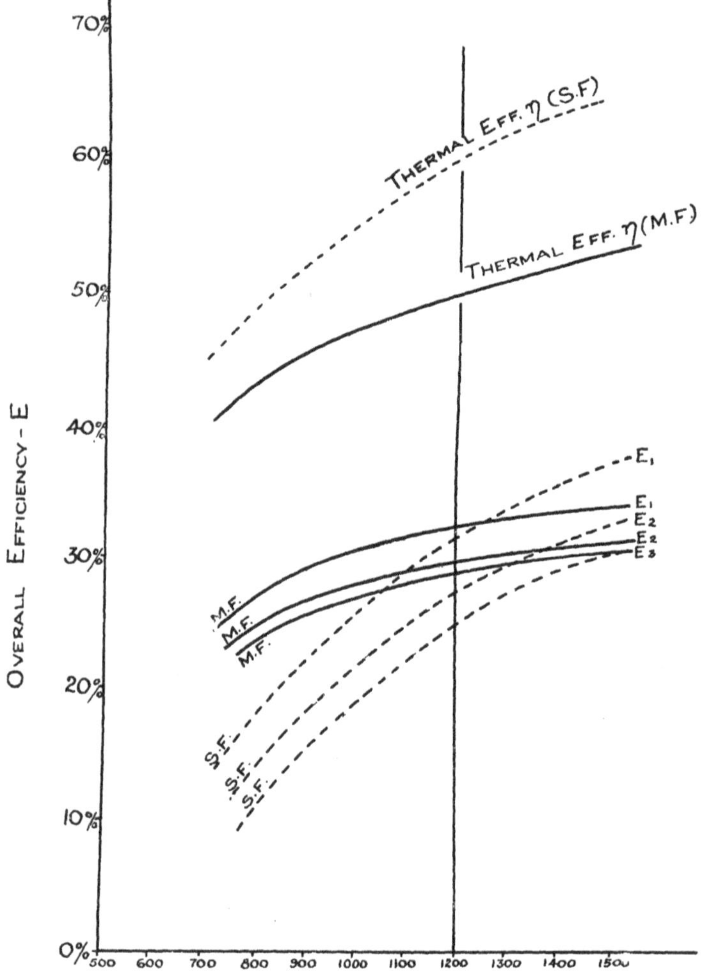

FIG. 34.—Temperature-Efficiency Diagram for Mixed-Fluid Turbine. E_1, Turbine, 75 per cent., Pump, 70 per cent. E_2, Turbine, 70 per cent.; Pump, 70 per cent. E_3, Turbine, 70 per cent.; Pump, 65 per cent.

70 per cent.; (2) turbine 70 per cent., pump 70 per cent.; (3) turbine 70 per cent., pump 65 per cent. A vertical line is

drawn through the locus of 1,200° C., as this represents what may be considered as the practical limit for initial temperature in a turbine driving on to the main shaft direct.

The dotted curves on the same diagram are reproduced from the efficiency-temperature diagram for the single-fluid constant-pressure turbine upon p. 37. It will be noticed that although the thermal efficiency of the latter type lies well above the thermal efficiency curve of the former, yet within practical limits of temperature the overall efficiency curves of the former type lie above the same curves for the latter.

For the thermodynamic efficiency of the turbine, 75 per cent., and of the pump, 70 per cent., the curves cut one another at 1,260° C. At 70 per cent. turbine and 70 per cent. pump they cut one another at 1,380° C., and at 70 per cent. turbine and 65 per cent. pump, at 1,500° C. Above these temperatures the single-fluid turbine is more efficient than the mixed-fluid turbine ; below these temperatures the reverse holds. These temperatures are, however, well above the limit of practical working ; in the case of the 75 per cent. turbine and 70 per cent. pump, the temperature at which the curves cut approaches the limiting value, but, in this instance, the efficiency taken for the turbine is somewhat high.

With 75 per cent. turbine and 70 per cent. pump, the mixed-fluid turbine shows a fuel consumption of 3·2 per cent. less than in the case of the single-fluid turbine for the maximum temperature limit—a saving of 8 per cent. when the thermodynamic efficiencies are each 70 per cent., and a saving of 16 per cent. when they are 70 per cent. and 65 per cent. It should also be noticed that the slope of the curves for the single-fluid turbine is much greater than that for the mixed-fluid turbines. Taking tangents to the two curves, for turbine efficiency and pump efficiency each 70 per cent.—

$$\frac{\delta E}{\delta \theta} = \cdot 214$$

for the mixed-fluid turbine,

$$\frac{\delta E}{\delta \theta} = \cdot 614$$

for the single-fluid turbine.

This is of considerable importance, as it means that there is a greater "flexibility" as regards efficiency variation in the mixed-fluid turbine, a fact that is of some importance when governing is effected by temperature control (*vide infra*). It should also be noted that the change in overall efficiency with

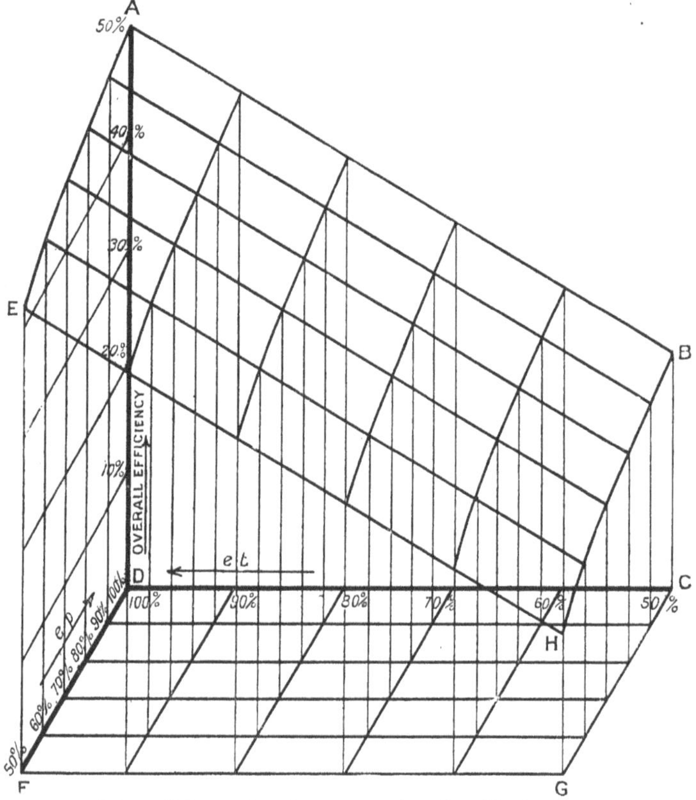

FIG. 35.—Three Dimension Diagram, showing Variation in Overall Efficiency, with Thermodynamic Efficiencies of Turbine and Pump. Thermal Efficiency assumed Constant at 50 per cent.

the change in thermodynamic efficiency is greater with the single-fluid turbine than it is with the mixed-fluid turbine. The only change in overall efficiency with variable factors that has not yet been analysed is the change in overall efficiency with the change in the thermodynamic efficiencies of the turbine and pump. As these two efficiencies are independently variable entities it is convenient to plot the change of overall efficiency with the variation of these in the form of a three dimension diagram, the change in the overall efficiency being represented by a surface (Fig. 35).

The thermal efficiency is fixed at 50 per cent., being the thermal efficiency corresponding to ten expansions in the turbine with an initial temperature of 1,117° C. The line AB in the plane ABCD represents the change in overall efficiency with the change in the thermodynamic efficiency of the turbine. The curve AE in the plane ADFE represents the change in the overall efficiency with the change in the thermodynamic efficiency of the pump.

The surface ABHE represents the change in the overall efficiency with the change in the efficiencies of the turbine and pump together. The overall efficiency at any point is given by the height of the vertical between the surface ABHE and the surface DCGF.

With the gas turbine that we have been considering there are three variable factors that determine the overall efficiency of the machine, viz., the initial temperature, the expansion ratio, and the thermodynamic efficiencies of the turbine and pump. Graphs have been drawn for the change of efficiency with each one of these variables, the other two being kept constant. It remains to be seen, however, whether each of these are independent variables. It is clear that the temperature and expansion ratio can be varied independently of one another. We have now to consider whether an increase in the initial temperature or an increase in the expansion ratio affects the value of the thermodynamic efficiency of the turbine.

Let it be assumed that the temperature of the working fluid after expansion—that is, the temperature on the blades—be fixed. Let the initial temperature then be varied, the expansion ratio undergoing alteration with it. It is to be seen whether the losses in the turbines are likely to be affected by this change of conditions.

The losses that are likely to occur in the turbine are as follows :—

(1) Residual velocity in working fluid after impact on the turbine wheel.

(2) Radiation losses through walls of turbine.

(3) Shock on vane edges.

(4) Shock of entrant fluid on quiescent fluid occupying vane passages.

(5) Spilling.

(6) Eddy currents.

(7) Friction in nozzle.

(8) Friction of the rotating wheel and blades.

(9) Friction of the working fluid through the blades.

(10) Friction of the turbine shaft in journals.

The first of these losses, it will be noticed, is controllable at will and is dependent only upon the peripheral speed at which it is decided to run the turbine wheel as compared to the velocity of the working fluid. With regard to the next loss, in the case of the steam-and-air turbine, in which the turbine itself is water-jacketed and serves in part as a steam generator for the raising of the dilutant steam, the radiation losses are negligible. With the single-fluid turbine, this is not so ; this matter will be considered later. The next three losses are dependent upon the construction of the turbine and may be assumed constant in any given machine. The loss due to eddy currents may also be considered as a constructional matter. The last four losses, however, are all frictional losses, and these are dependent upon the surface and the velocity of the moving bodies within or upon which the friction occurs. The last item is a relatively small one, and except to note that it is dependent on the velocity of the turbine shaft, and therefore upon the velocity of the working fluid, it may be ignored. The other three frictional losses, it will be noticed, are all losses due to the friction of the working fluid over metal surfaces. This loss has been shown by Odell, Stodola and others to be dependent upon the velocity of the gas (or body in the gas), the density of the gas and the surface of the body. This latter with any particular turbine is fixed. Further, it has been shown that the loss due to this friction—

$$\delta = k \,.\, v^3 \,.\, \mathrm{S} \,.\, \rho$$

where k is a constant, v is the relative velocity between the surface and the gas, S is the area of the surface, and ρ is the density of the gaseous medium. Now the velocity of the fluid on expansion is dependent on the initial temperature, the final temperature being fixed, the relation between them being given by the equation —

$$v^2 = k_1 \,.\, (\theta_1 - \theta_2)$$

k_1 being a constant.

Therefore, we have—

$$\delta = k_2\, (\theta_1 - \theta_2)^{\frac{3}{2}} \,.\, \mathrm{S} \,.\, \rho$$

but, in any one machine, s, the rotor surface, is constant; and as θ_2 is fixed, ρ, the density of the medium, is also constant. Therefore—

$$\delta = \mathrm{K}\, (\theta_1 - \theta_2)^{\frac{3}{2}}$$

Now, work done on expansion—

$$\mathrm{W} = \mathrm{C}_p\, (\theta_1 - \theta_2)$$

Therefore, percentage of work lost in friction—

$$x = \frac{\delta}{\mathrm{C}_p\, (\theta_1 - \theta_2)}$$

$$= \frac{\mathrm{K}}{\mathrm{C}_p} \sqrt{\theta_1 - \theta_2}$$

$$= \mu \sqrt{\theta_1 - \theta_2}$$

μ being a constant.

In the diagram (Fig. 36) is plotted in the dotted curve, KL, the relation, $x = \mu \sqrt{\theta_1 - \theta_2}$, in the form of percentage loss due to friction against initial temperature, θ_2 being assumed constant at 500° C., and δ to be equal to 10 per cent. of the "indicated" work at 800° C. This diagram and the succeeding one are drawn for the single-fluid turbine only.

It will be seen that an increase in the initial temperature of the system demands an increased ratio of expansion, as the temperature after expansion, θ_2, remains constant. That is to say, the negative work is increased, a higher compression has to be effected, and the pump, therefore, has to run at an increased peripheral speed. Thus, in considering the effect of frictional losses with rise in initial temperature, it must be borne in mind that an increase in frictional loss occurs both in the positive turbine unit and in the negative pump unit.

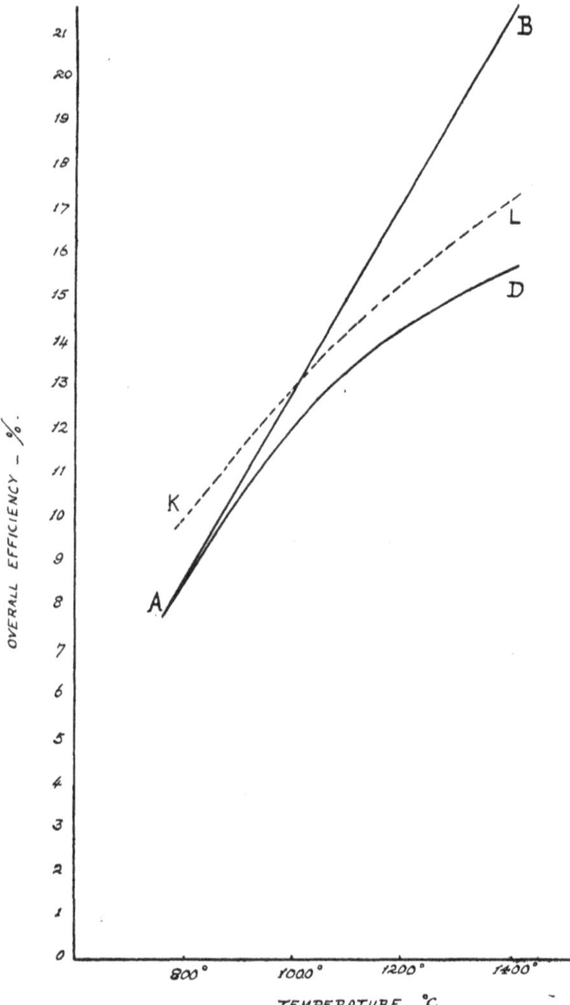

FIG. 36.—Diagram showing Change in Overall Efficiency with Rise in Initial Temperature, taking into account Frictional Losses. (Single-Fluid Turbine.) $\delta = 10$ per cent. W_0 at 800° C.

Now, the work done by the pump—

$$\omega = p_0 v_0 \log_\epsilon r$$

where r is the compression ratio.

Since, however, the compression ratio in the pump is the same as that in the turbine—

$$r = \left(\frac{\theta_1}{\theta_2}\right)^{\frac{\gamma}{\gamma - 1}}$$

and—

$$\omega = p_0 v_0 \log_e \left(\frac{\theta_1}{\theta_2}\right)^{\frac{\gamma}{\gamma - 1}}$$

we may take it that the peripheral velocity of the pump rotor is proportional to the square root of the work done ; or—

$$v = k_3 \sqrt{\omega}$$

but the frictional loss in the pump, δ_p, is proportional to the cube of the velocity. Therefore—

$$\delta_p = k_4 \, \omega^{\frac{3}{2}}$$

Let it be assumed that at an initial temperature of 800° C. the total losses in the turbine amount to 30 per cent. of the indicated work, and that 10 per cent. of the indicated work is loss due to friction. Let it be assumed, also, that in the rotary compressor the losses amount to 30 per cent., and that two-thirds of this loss is caused by disc friction. Let W be the work done by the turbine, and ω the work done by the pump. Then, the overall efficiency—

$$E = \frac{(\cdot 8W - \delta) - \left(\frac{\omega}{\cdot 9} + \delta_p\right)}{W + h}$$

where h is the heat lost from the regenerator ; taking the exit temperature of the regenerator at 215° C., $h = 50$ T.U.

In Fig. 36 the graph AB shows the change in overall efficiency with δ and δ_p assumed constant. The curve AD shows the change with δ and δ_p varying in accordance with the formula quoted above.

There is another factor, however, which affects the overall efficiency with rise in initial temperature which has not yet been taken into account. This is the heat loss due to radiation from the combustion chamber and other parts of the fluid path. This loss is proportional to the fourth power of the absolute temperature, or—

$$\delta' = k_5 . \theta_1^4.$$

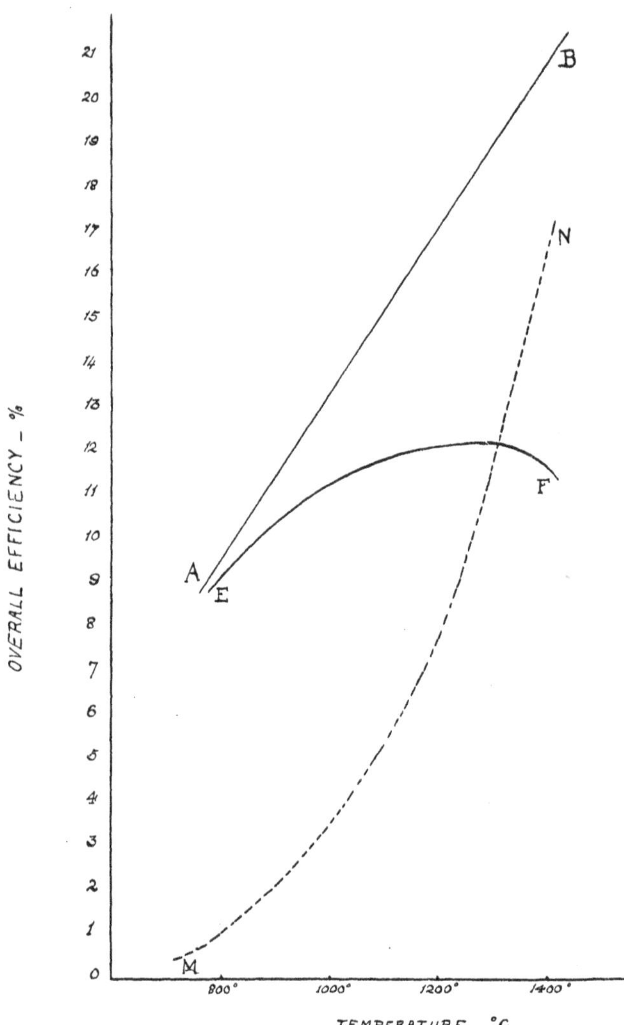

FIG. 37.—Diagram showing Change in Overall Efficiency with Rise in Initial Temperature, taking into account both Frictional and Radiation Losses. (Single Fluid Turbine.) $\delta = 10$ per cent. W_0 at 800° C.; $\delta' = 6$ per cent. W_0 at 800° C.

This relation is shown by the dotted curve MN (Fig. 37). The graph AB in Fig. 37 is the same as the graph AB in

Fig. 36. Taking the heat loss δ' into account, in the equation for overall efficiency we have—

$$E = \frac{(\cdot 8W - \delta) - \left(\frac{\omega}{\cdot 9} + \delta_p\right)}{W + h + \delta'}$$

This relation is plotted in the curve EF. It will be seen that a point is eventually reached at which any increase in the initial temperature merely produces a lowering of the overall efficiency of the machine.

From the above review of the possibilities of the mixed-fluid constant-pressure gas turbine the following generalisations may be inferred. First, that neither with the single-fluid turbine nor with the mixed-fluid turbine is it advantageous or economical to work beyond a given limit of temperature and expansion ratio; and that these limits are probably below those fixed by constructional limitations. Secondly, that with the mixed-fluid turbine the most efficient value for the expansion ratio lies under ten, and therefore it is obviously more economical, as well as more convenient on every count, to work the turbine entirely below the atmosphere, more especially as the frictional losses in both turbine and pump vary directly with the density of the medium in which the rotors rotate. And thirdly, that in the case of the mixed-fluid turbine the troublesome question of radiation losses is avoided, and that, as the turbine in question is of necessity sub-atmospheric, it may be possible to use the gases from the producer (in the case of gaseous fuel) without previous cooling and so save something of the 16 to 20 per cent. producer loss which is largely due to the initial cooling of the gas. This matter is discussed more fully in Chapter VII.

There is one important disadvantage possessed, or likely to be possessed, by the steam-and-gas turbine, and that is the possible formation of sulphuric acid brought about by the reaction of the sulphur contained in the gas with the water with which the gas is mixed. This question, together with the methods by which it may be possible to counteract the evil, is considered separately, in Chapter VII. This difficulty also occurs in some cases with oil fuels. More is to be said of the part played by oil fuels in the question of the gas turbine later; it need not be discussed here. It might, however, be not out of place to convert the overall efficiency of the steam-and-air turbine, described on pp. 61—66, into lbs. of oil burnt per horse-power per hour. The overall efficiency in question amounted to 28·5 per cent. Let the thermal value of the oil be taken at 10,000 T.U. per lb. Then, assuming the furnace losses to be *nil*, the consumption of oil equals ·5 lb. per brake horse-power per hour. Assuming there to be a loss in the furnace of 5 per cent., and it must be remembered that, owing to the water-jacketing, this loss must of necessity be small, the oil consumption is raised to ·523 lb. per brake horse-power per hour.

This concludes the general survey of the constant-pressure gas turbines. Certain evolved types of mixed-fluid turbines will be considered in the next chapter, and a more detailed analysis of the construction and working, and of the accessory machinery to these types of gas turbines, will be considered in Book II.

CHAPTER IV. MIXED-FLUID TURBINES: DERIVED TYPES

In the last chapter the question of the addition of steam to the constant-pressure, internally-fired gas turbine was investigated. In the type therein discussed the steam was not only added to the system but actually mixed with the products of combustion themselves, a composite working fluid being thus produced which was expanded through the turbine. The application of steam (or other condensible fluid) to other types of constant-pressure gas turbines is to be now considered, as well as that particular case of the mixed-fluid constant-pressure gas turbine in which the steam is expanded separately—wholly or in part—from the products of combustion.

It is, of course, possible to conceive a constant volume or explosion gas turbine working with the addition of steam—either as a constituent of the working fluid or expanded separately as an adjunct to the system. As a matter of fact, this latter is actually what is done in the Holzwarth turbine—the only explosion turbine that has given anything approaching reasonable economy in actual trial. In the turbine in question the waste heat from the gases after expansion on the turbine wheel are led into a regenerator, where the waste heat is abstracted in raising steam, which is used on a separate turbine wheel to drive the pump that undertakes the initial compression of the explosive mixture. It is, however, not convenient to treat such under the head of mixed-fluid turbines. It is considered separately in the following chapter as a simple constant-volume gas turbine. The addition of steam to the explosion turbine has really only an accidental and not an absolute value. It may be convenient so to do; it is not efficient so to do, for the negative work in turbines of this type is absent (Lenoir type—Karavodine turbine) or else small compared to the positive work (Otto type—Holzwarth turbine) and the addition of steam does not minimise the percentage of negative work sufficiently to compensate for the inferior thermal efficiency of the steam.

The derived types of mixed-fluid turbines other than the type considered in the preceding chapter may be examined under the four following heads :—

I. The application of steam to the externally-fired constant-pressure turbine.

II. The application of steam to the internally-fired constant-pressure turbine in which steam expansion takes place on separate wheels.

(*a*) Steam expanded separately in whole.
(*b*) Steam expanded separately in part.

III. The approach to the ideal of isothermal expansion: interheating. (*Note.*—This hardly falls under the head of "derived mixed-fluid systems," but it is a mixed-fluid turbine type and may be conveniently treated here.)

IV. Mixed-fluid turbines using other condensible fluids than steam. That is SO_2, etc.

Of these four types, the only ones that afforded efficiencies consonant with those of other prime movers are types II. (*b*) and III. Types I., II. (*a*), and IV. are of theoretical interest only.

I.—THE APPLICATION OF STEAM TO THE EXTERNALLY-FIRED CONSTANT-PRESSURE TURBINE.

In considering the externally-fired gas turbine it was seen that the type in which the cycle was "open," the expansion super-atmospheric and the regeneration complete, approached nearer practicability in working than any other kind. The restricting of the upper temperature limit, however, to that value necessitated by the resisting power of the tubes used in the furnace, caused the negative work to bulk too largely in the system and to produce a consequently low efficiency. If, however, part of the positive work is undertaken by steam, the negative work becomes a less percentage of the positive work, and an increase in efficiency is obtained. The question is analysed in the following manner.

Let the maximum temperature of the air used as the working fluid be 500° C.; let a temperature difference of 150° C. be required to effect the flow of heat through the furnace tubes. This makes the temperature of the gases entering the furnace 650° C. and the temperature of those same gases leaving the furnace 165° C. The combustion mixture entering the furnace is assumed to be of a thermal value of 550 T.U. per lb. The temperature of the heating fluid is lowered to a temperature of 650° C., at which it enters the tubular heater by raising steam. Heat is further abstracted from the furnace gases after passing through the air heater by means of a feed-water heater; the temperature therein being lowered from 165° C. to 115° C.

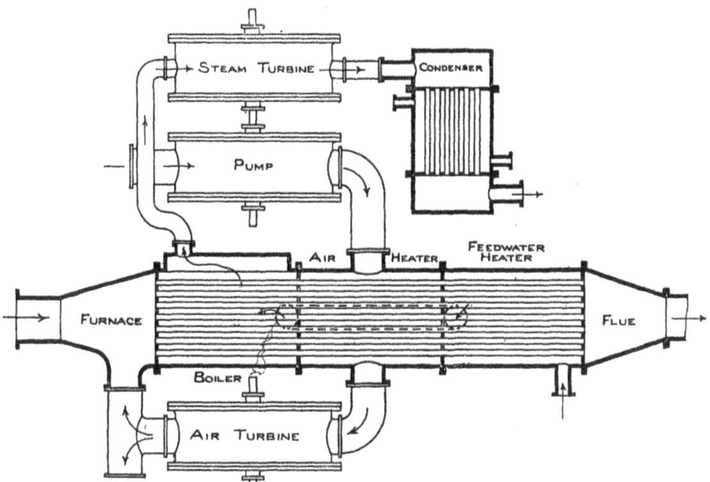

Fig. 38.—Externally-Fired, Constant-Pressure, Mixed-Fluid Turbine.

The steam raised by the surplus heat of the furnace gases is expanded on a turbine wheel. The air passing through the heater is initially compressed by a rotary pump, expanded in the air turbine, and exhausted direct into the furnace. A diagrammatic representation of the system is shown in Fig. 38.

Let it be assumed that the steam added is expanded from 150 lbs. to 1 lb. The thermal efficiency of steam expanded between these limits is 28·3 per cent. If the efficiencies of the turbine and boiler be each taken at 70 per cent., the overall efficiency of a steam turbine installation under these conditions is 13·9 per cent. The steam adjunct to this air turbine is in the same category as the above steam turbine and boiler plant;

the steam and air turbine under consideration must, then, give a final overall efficiency greater than 13·9 per cent., greater, that is, than the efficiency of the steam alone, if the addition of the steam is to be of any use. The final overall efficiency of the combined turbine must also be greater than the efficiency of the turbine under the same conditions, provided no steam had been added. As, however, without the addition of the steam the air turbine would possess a *negative* efficiency (*vide* p. 24), that is, that the turbine would not even turn round, there is plenty of surety for the steam addition on this count.

Let the thermal value of the products of combustion be, as before, 550 T.U. per lb.

Then the heat given up to the air passing through the heater—

$$H_1 = C_p\,(650 - 165) = 121\ \text{T.U.}$$

The furnace gases leave the feed-water heater at a temperature of 115° C. Therefore heat to waste equals 25 T.U. Therefore, heat left to raise steam—

$$H_2 = 550 - 121 - 25 = 404\ \text{T.U.}$$

Taking it that, approximately, 1 lb. of air is required for the combustion of 1 lb. of gas, one half of the exhaust from the air turbine may be used for this purpose, and all the residual heat in that half may be recovered ; it may be assumed that the other half of the heat of the exhaust goes to waste.

Let the number of the expansions in the furnace be six. The temperature then of the exhaust from the turbine would be 197° C. This would give a heat recovery of 23 T.U.

The work done by 1 lb. of steam from 150 lbs. to 1 lb. is 176 T.U. The heat required to raise this steam is 623 T.U. per lb.

Therefore quantity of steam raised is—

$$\frac{404}{623} = {\cdot}65\ \text{lb.}$$

Therefore the work done by the steam is 114 T.U. The work done by the air is 76 T.U. This gives a total positive work of 190 T.U.

Negative work—

$$W_n = \frac{p_0 v_0}{J} \log^e r = 18{\cdot}7 \log_e 6 = 33{\cdot}5\ \text{T.U.}$$

Nett heat put into system—

$$H = 550 - 23 = 527 \text{ T.U.}$$

Therefore

$$\eta = \frac{190 - 33{\cdot}5}{527} = 29{\cdot}7 \text{ per cent.}$$

Taking the value of the thermodynamic efficiencies of turbine and pump to be each 70 per cent., we have a final value for the overall efficiency—

$$E = 16{\cdot}1 \text{ per cent.}$$

This is far too low a value to stand in competition with overall efficiencies given by other types of gas turbines and prime movers, but it is interesting to note the rather paradoxical result that by combining with the steam turbine a type of gas turbine giving by itself a negative efficiency, the final overall efficiency of the combined systems is higher than that of the steam plant taken by itself.

II. THE APPLICATION OF STEAM TO THE INTERNALLY-FIRED CONSTANT-PRESSURE GAS TURBINE IN WHICH THE STEAM EXPANSION TAKES PLACE ON SEPARATE WHEELS.

(a) Steam Expanded Separately in Whole.

Suppose that in an ordinary single-fluid constant-pressure turbine the pump, instead of being driven by a gas turbine unit, is driven by a steam turbine. Let this steam power unit be entirely independent of the gas turbine, and let the steam expand between the pressures of 150 lbs. and 1 lb. This gives a thermal efficiency of 28·3 per cent. If the boiler and turbine be each taken at 70 per cent. efficiency, we have, as before, an overall efficiency for the plant of 13·9 per cent.

Let the expansion ratio in the gas turbine be ten, and the initial temperature be 1,200° C. Then positive work done by the products of combustion equals 172 T.U.

The work done by the pump per lb. of fluid pumped is 43 T.U. This is driven by the steam turbine. The pump thermodynamic efficiency is 70 per cent., the overall efficiency of the system driving it 13·9 per cent. Therefore, heat expended in driving the pump—

$$h = \frac{43}{\cdot 7 \times \cdot 139}.$$

Therefore—

$$h = 442 \text{ T.U.}$$

This may be regarded simply as a heat loss.

Let the regeneration be calculated on the same basis as that for the single-fluid constant-pressure turbine discussed in Chapter II. (*vide* p. 33). Then heat lost in waste gases equals 50 T.U. per lb. of fluid. Therefore total heat put into system—

$$\text{H} = 172 + 50 + 442 = 664 \text{ T.U.}$$

Work done by gas turbine at 70 per cent. thermodynamic efficiency equals 120 T.U. Therefore overall efficiency of the system—

$$\text{E} = 120/664 = 18 \text{ per cent.}$$

Here, again, it will be noticed that the overall efficiency is not as great as it is in the preceding types discussed. The deficiency is more easily explainable when it is seen that the thermal efficiency of steam expanded from 150 lbs. to 1 lb. without superheat is 28·3 per cent., while in the case of steam expanded from 15 lbs. to 1 lb. and *superheated* to a temperature of 1,117° C. the thermal efficiency is 31·2 per cent., and also that the 30 per cent. boiler loss which occurs in the first case is avoided in the latter.

It may also be noticed that the above system is parallel to that of Holzwarth, the only difference being that, in the latter case, the heat is introduced into the working fluid of the gas turbine at constant volume instead of at constant pressure.

A diagrammatic illustration of the constant-pressure, internally-fired gas turbine, with a steam-driven pump, is shown in Fig. 39.

(b) Steam Expanded Separately in Part.

In the sub-atmospheric type of mixed-fluid turbine considered in Chapter III. the steam that is added to the products of combustion is at atmospheric pressure during evaporation in the furnace and regenerator. The pressure, however, of the steam in the furnace-jacket and in the regenerator (*vide* Fig. 29) is, in actual fact, arbitrary. The steam may be evaporated at a pressure above the atmosphere, and then, after expansion to the atmospheric pressure through a suitable wheel, be added to the combustion products in the furnace.

A turbine so designed is clearly an evolved type from the

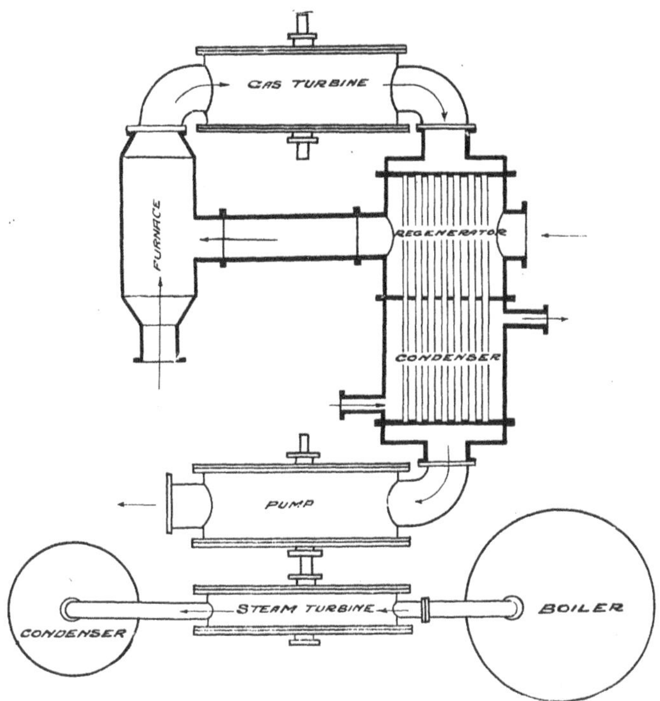

FIG. 39.—Internally-Fired, Constant-Pressure, Mixed-Fluid Turbine. Air Turbine, Sub-Atmospheric. Steam expanded separately throughout the whole Cycle.

sub-atmospheric turbine of Chapter III. Indeed, the turbine can be so designed as to admit of the use of a high-pressure steam wheel or not, as the user of the plant may determine. The system of partial separate steam expansion is shown diagrammatically in Fig. 40. The products of combustion evaporate steam in a high-pressure boiler—or the equivalent of such—and the steam thus generated is expanded down to the atmospheric pressure in a suitable turbine. The exhaust from this turbine mingles with the products of combustion from the boiler-flue and the composite fluid thus produced expanded down to the terminal pressure, the requisite difference in pressure being maintained by a rotary pump. The exhaust gases from the low-pressure turbine pass through a regenerator, from which the high-pressure boiler is fed.

A system of this nature the author believes to be the most efficient possible among gas turbines of all denominations within the limits of pressure, temperature and expansion ratio (and even beyond these limits) set by practice.

The efficiencies of this type of turbine may be estimated on

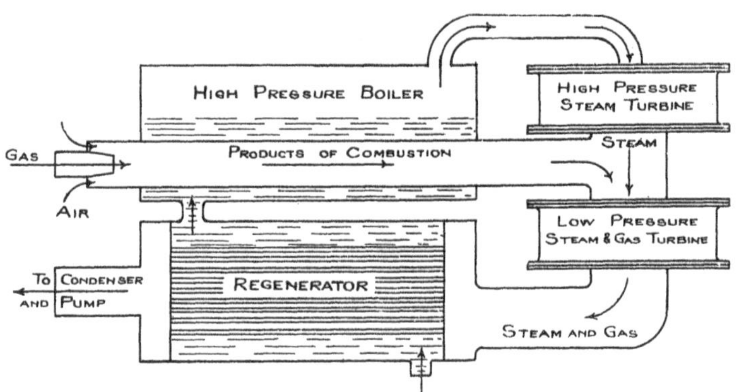

FIG. 40.—Internally-Fired, Constant-Pressure, Mixed-Fluid Turbine. Steam expanded separately only in high-pressure stage.

the same basis as those of the sub-atmospheric, constant-pressure, internally-fired, mixed-fluid turbine.

Let the pressure in the high-pressure boiler be 150 lbs. per square inch. Let the low-pressure steam and gas turbine work between the pressure limits of 15 lbs. and 1·5 lbs. Let the initial temperature in the furnace be, as before (*vide* p. 56), 1,390° A.

The Quantity of Steam Added.—This is found by the usual formula—

$$y \,.\, m = \mathrm{H} - \mathrm{C}_{pa}(\theta_1 - \theta_0) + (t_1 - t_2)(\mathrm{C}_{pa} + \mathrm{C}_{ps} \,.\, y).$$

m here equals the quantity of heat required to raise 1 lb. of steam from the condenser temperature to the maximum temperature and pressure. This gives a value for m of 1,101 T.U. Taking the value for other constants in the equation, as given on p. 57, the value of y becomes ·41 lb. The work done by 1 lb. of steam expanded from 150 lbs. down to 15 lbs. without superheat is 93 T.U. The work done by 1 lb. of steam expanded from 15 lbs. to 1·5 lbs. superheated to 1,390° A. is 352 T.U. Work done by 1 lb. of air (*vide* p. 57) is 172 T.U. and the negative work on the pump 43 T.U. Then total positive work in the system is—

$$
\begin{aligned}
W &= \cdot 41\,(93 + 352) + 172\,.\,\text{T.U.} \\
&= 354\ \text{T.U.}
\end{aligned}
$$

The total heat put into the system is 550 T.U.

Therefore—

$$\eta = \frac{354 - 43}{550} = 56{\cdot}6 \text{ per cent.}$$

and the overall efficiency, taking the thermodynamic efficiencies of the pump and turbine to be each 70 per cent.—

$$E = 34 \text{ per cent.}$$

This is the greatest efficiency that we have so far found to be attainable within those definite limits that have been adopted as coincident with those of practice. It signifies a saving of 16·2 per cent. in fuel consumption on the mixed-fluid turbine of Chapter III., and shows a fuel economy equal, if not superior, to that of the best Diesel engines.

This overall efficiency of 34 per cent. represents, if oil fuel be used, a consumption per brake horse-power per hour of ·414 lb. of oil, taking the calorific value of the oil to be 10,000 T.U. per lb. If the thermodynamic efficiency of the turbine can be raised to 75 per cent., the overall efficiency becomes 37·5 per cent. and the consumption of oil ·376 lb. per brake horse-power. This is a less fuel consumption than any known prime mover in actual work.

Taking coal at a calorific value of 750 T.U. per lb., and a gas producer efficiency of 84 per cent. (Mond Gas Producer), the fuel consumption in the case, E = 34 per cent., becomes ·65 lb. of coal, and with E = 37·5 per cent., ·59 lb. In the case of the sub-atmospheric turbine working between the same limits, but without the high-pressure steam wheel, the fuel

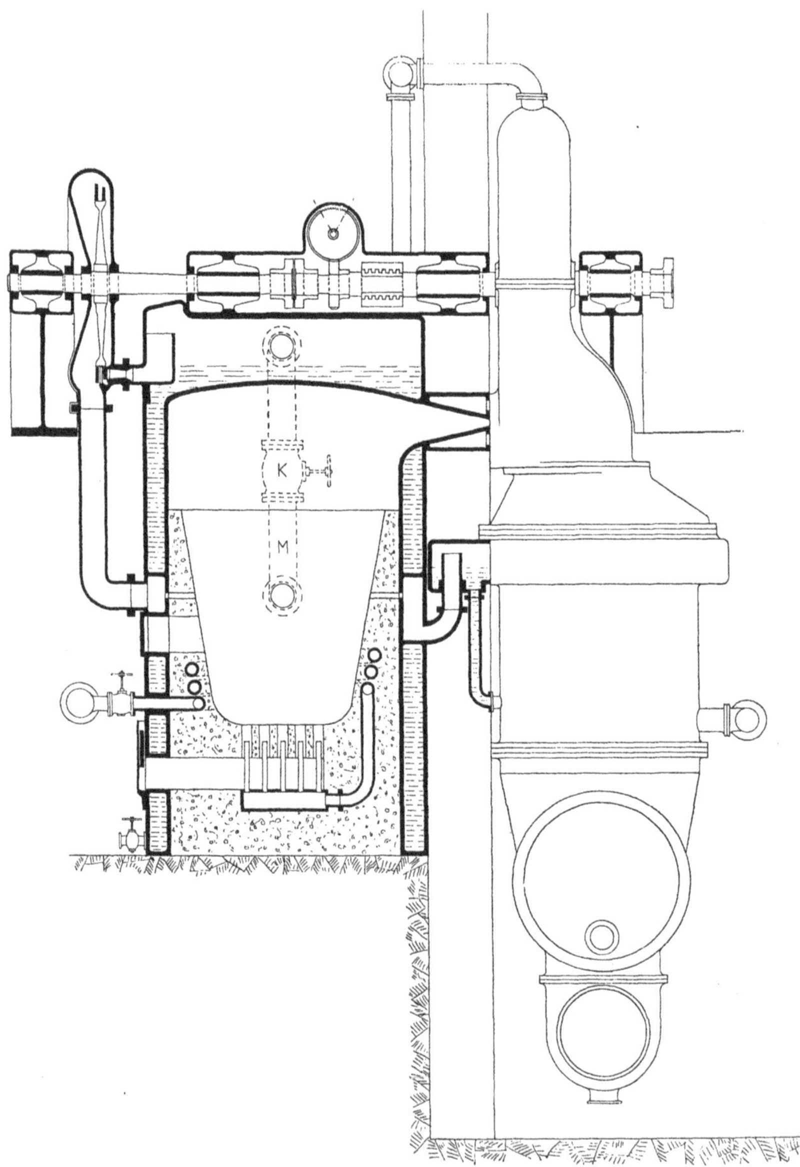

Fig. 41.—Davey Mixed-Fluid Turbine, with High-Pressure Steam Wheel. Scale $\frac{1}{4}$ inch to one Foot.

consumption is ·77 lb. of coal or ·5 of oil (assuming the thermodynamic efficiencies of the turbine and pump to be each 70 per cent.).

The application of the high-pressure steam wheel to the sub-atmospheric turbine described in Chapter III. is shown in Fig. 41. The design is essentially the same as that shown in Fig. 29. The steam wheel, however, is so arranged as to allow of uncoupling from the main turbine shaft. The steam from the furnace jacket can then be bye-passed direct into the furnace and the turbine run sub-atmospherically, without the use of the steam wheel. The steam wheel is placed close to the furnace at the back of the main turbine. The coupling, necessary bearings, thrust-block and governor pinion, are all carried above the boiler, between the two turbine wheels; in this manner considerable space is saved at the driving end of the turbine. The rest of the design is precisely the same as in the low-pressure turbine, the furnace and regenerator merely being made of that strength necessary to withstand the pressures adopted.

The steam wheel, of course, becomes smaller in diameter than the main turbine wheel. The velocity of the steam—

$$v = \sqrt{W.2g.J.}$$

W equals 93 T.U. and this gives a value for v of 2,900 ft./sec. compared to the velocity of the gases on the main turbine wheel of 4,490 ft./sec. The addition of the steam wheel also affords a ready and sensitive means of governing; this will be more fully discussed in Chapter VII.

III. THE APPROACH TO THE IDEAL OF ISOTHERMAL EXPANSION: INTERHEATING.

If, in the constant-pressure, single-fluid gas turbine of the internally-fired type, the expansion as well as the compression were to take place isothermally, the cycle known as the Erricson cycle, and referred to in Chapter II., would be secured. This is a heat cycle that gives the same thermal efficiency as the Carnot, viz., $\frac{T_1 - T_2}{T_1}$ equals η.

As has been already pointed out, the achievement of isothermal expansion is not permissible. One step in that direction can, however, be secured, and a condition of expansion brought about between that of the true isothermal and true adiabatic transformations. This is achieved by a system of reheating the working fluid after it has undergone a certain temperature drop due to adiabatic expansion. The oftener the reheating takes place during the adiabatic expansion of the fluid, the more nearly does that expansion approach to the isothermal; in the limit the expansion, of course, becomes isothermal.

Fig. 42 shows this effect of reheating on the pressure-volume diagram. The "PV" diagram is drawn for ten expansions,

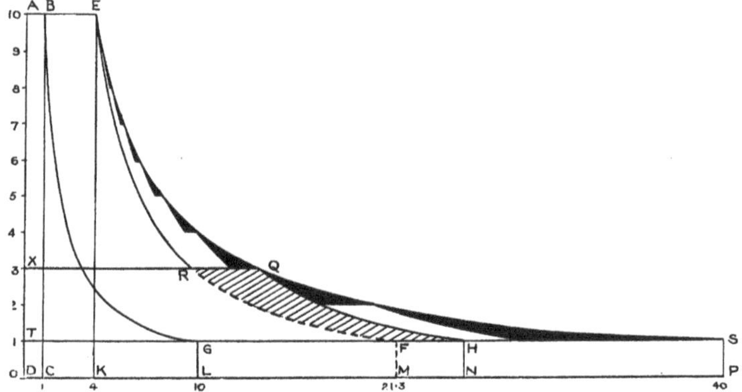

FIG. 42.—"PV" Diagram, showing Effects of Reheating.

with a value for γ of 1·38, and is similar to the diagram in Fig. 10. The curve ES represents the isothermal expansion of the gas, the curve EF the adiabatic expansion. The curve GB represents the isothermal compression of the pump. If isothermal expansion were complete, the work done by the gas would be represented by the area AEST; if the expansion were wholly adiabatic, by the area AEFT. Suppose the gas to expand adiabatically for every lb. drop in pressure, but that at the end of each lb. drop heat be introduced from some external source to raise the temperature of the gas to the initial value. The work done by the gas would then be represented by the area AEST *minus those areas shown black on the diagram.* Imagine, however, that heat be introduced at one point only during the adiabatic expansion of the gas, for instance, when the gas has expanded from 10 to 3. The gas being then raised to the initial temperature the volume is increased from XR to XQ; further adiabatic expansion then occurs along QH. The work done in this instance is represented by the area AERQHT; and extra work gained by the reheating by the shaded area RQHF. Fig. 43 is a temperature-entropy diagram showing these same effects of reheating. The work done with complete isothermal expansion is represented by the area AQBK; that with adiabatic expansion by the area

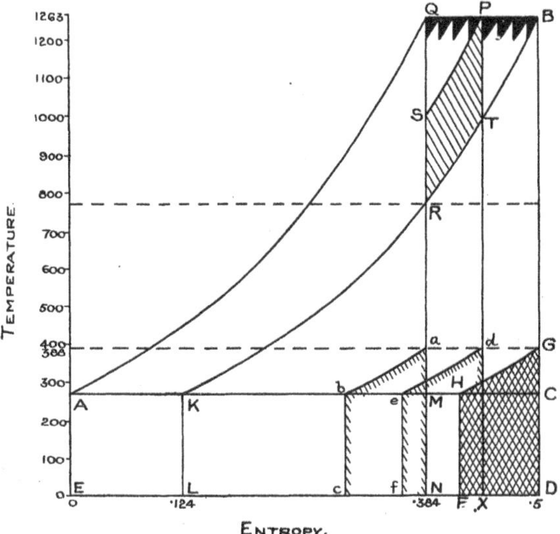

FIG. 43.—"ΘΦ" Diagram, showing the Effects of Reheating.

AQRK. If the reheating is frequent, the work done is represented by the area AQBK, *minus the portions of the diagram shown in black.* If the reheating takes place at one point only, the work done is represented by the area AQSPTK, and the extra work added by the process by the shaded area SPTR.

The extra heat put into the system is, of course, shown by the areas lying beneath the curves SP and RB, or the horizontal QB.

At first there would seem to be a loss in the process, owing to the fact that the quantity of added heat, QPXN, is utilised at a less efficiency than the original quantity of heat, AQNE.

This would be the case were there no regeneration. But with regeneration the quantity of heat to waste remains constant (theoretically) whether heat be introduced during adiabatic expansion or no. In the case of the steam-and-gas turbine, using a steam regenerator, the exit temperature of the waste gases was computed at 115° C. This loss of heat is shown in the diagram by the areas GHFD, *def*X, *abc*N, according as the expansion is isothermal (or approximately so), receptive of one reheating, or wholly adiabatic. The areas are equal. Fig. 43 is drawn for the same conditions as Fig. 11 (*q.v.*).

The effect of reheating with superheated steam is similarly shown in the temperature-entropy diagram for steam, Fig. 44.

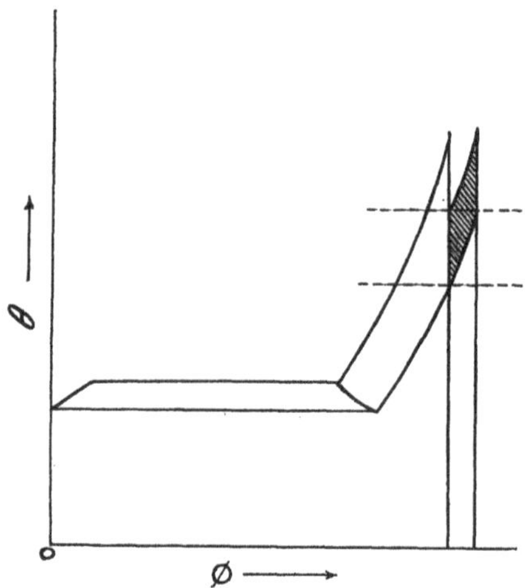

FIG. 44.—"ΘΦ" Diagram for Steam, showing the Effect of Reheating.

The scheme of reheating the working fluid during adiabatic expansion is by no means new, and a patent (No. 24,781) was taken out in 1902 by de Ferranti. The proposals therein made referred rather to the steam turbine than to the gas turbine, and the reheating was frequent. The steam was led through tubular heaters between the turbine elements. In this manner a condition approaching that of isothermal expansion was attained. Whether such a system of reheating through the medium of tubes is likely to be at all practicable with regard to steam may be doubtful, but there is certainly no doubt but that the process is inadmissible with the gas turbine. The lower limit of temperature of the fluid is, in the gas turbine, about 500° C., and to raise gases at such a temperature to any higher value would be impossible by means of a tubular heater—the temperatures involved being too great for the resisting power of the metal. If, therefore, there is to be any reheating of the working fluid, it must be effected by the internal, and not by the external, application of heat. The only efficient way in which heat can be introduced into a system by internal application is by the introduction of the fuel at atmospheric pressure into a system working with an expansion range both above and below the atmospheric pressure. In this instance no loss is experienced owing to the necessity for initial compression of the added fluid, as would be the case were the addition to occur at any other pressure than that of the atmosphere.

It is thus plain that only one operation of reheating can take place, viz., when the working fluid has been expanded down to the atmospheric pressure.

Let it be assumed that the same expansion occurs above the atmosphere as below. It will be observed that in a system of this sort, the pumping plant has to be in duplicate, one pump acting as compressor, the other as exhauster. As additional fluid is introduced into the system at atmospheric pressure, the exhausting pump must be of an increased capacity. Let the total range of the expansion be twenty-five expansions, five above and five below the atmosphere. To attempt larger expansion ratios than this would incur the use of pressures too high for the efficient working of a rotary pump. The consideration of the steam-and-gas turbine under these conditions may be undertaken in the following manner.

As before, let the temperature of the gases after expansion be 500° C. This is the temperature of the gases at atmospheric pressure after the expansion from 75 lbs. absolute. Assuming the value of γ for the products of combustion to be 1·38 and that for steam to be 1·3, and taking an arbitrary value for the quantity of steam added at ·5 lb. (in reality ·606 lb.) we get an approximately correct value for the initial temperature of 1,180° A. or 907° C.

The work done by the steam in expanding from 75 lbs. to 15 lbs., at a superheat of 907° C., is 254 T.U., and the work done by the steam from 15 lbs. to 1·5 lbs. with the same superheat is 269 T.U. Therefore, the total work done by 1 lb. of steam throughout the system equals 523 T.U. The work done by 1 lb. of the products of combustion in each half of the expansion range is 105 T.U.

The Quantity of Steam added.—The formula used above

(*vide* p. 57) to obtain the value of the weight of steam added will not be permissible here, for there is an unknown quantity in the weight of fuel added after the initial expansion has taken place. Let this value be x.

Then, we have—

$$y \,.\, m = \mathrm{H} - \mathrm{C}_{pa}(\theta_1 - \theta_0) + (t_1 - t_2)(\mathrm{C}_{pa}(1 + x) + \mathrm{C}_{ps} \,.\, y).$$

Substituting the values found above, we have—

$$y = {\cdot}53 + {\cdot}11x.$$

To find the value of x we have the equation—

$$\mathrm{H} \,.\, x - \mathrm{C}_{pa}(\theta_1 - \theta_0)\, x = (\theta_1 - \theta_2)(\mathrm{C}_{pa} + y \,.\, \mathrm{C}_{ps}).$$

Substituting the values as before, we have—

$$y = 1{\cdot}6x - {\cdot}5.$$

We thus have a simultaneous equation for the values of x and y, giving—

$$x = {\cdot}690$$

and

$$y = {\cdot}606.$$

Therefore, total negative work done on pump—

$$\mathrm{W}_n = pv \,.\, \log_e {\cdot}5 \times 2{\cdot}69$$
$$= 81 \text{ T.U.}$$

Total positive work done by the air—

$$\mathrm{W}_p = 105 \times 2{\cdot}69 = 283 \text{ T.U.}$$

The total work done by the steam in the system—

$$\mathrm{W}_s = 523 \times {\cdot}606 = 320 \text{ T.U.}.$$

Therefore total positive work = 603 T.U.

" " negative work = 81 T.U.

" " heat put in = $550 \times 1{\cdot}69 = 930$ T.U.

Therefore

$$\eta = \frac{603 - 81}{930} = 56 \text{ per cent.}$$

And the overall efficiency, taking $e_t = e_p = {\cdot}7$—

$$\mathrm{E} = 33 \text{ per cent.}$$

It will be seen that here the overall efficiency obtained closely approaches that of the steam-and-gas turbine with the high-pressure steam wheel. The system has, however, certain disadvantages.

It may be asked, if the system of reheating is effective in this case of the steam-and-air turbine, would it not be even more effective in the case of the single-fluid, constant-pressure gas turbine? In point of fact, this is not the case. Owing to the confinement of the expansion range to five in the initial portion of the turbine, the maximum temperature is limited to 927° C., taking the temperature on the blades of the turbine to be 500° C. The overall efficiency ($e_t = e_p = \cdot 7$) of the single-fluid gas turbine without the reheating, at an initial temperature of

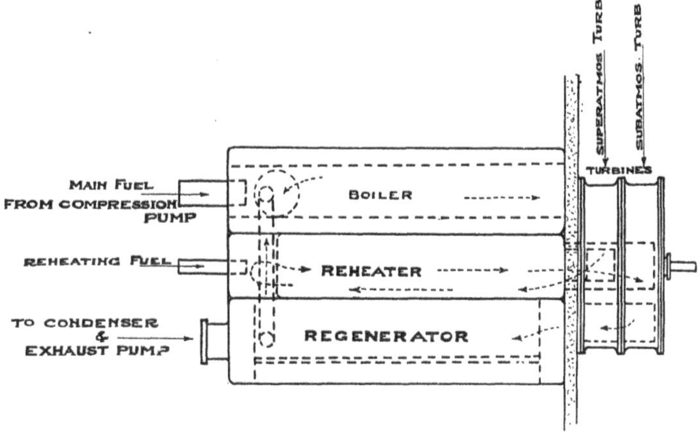

FIG. 45.—Diagrammatic Representation of Steam-and-Air Turbine with Reheater.

927° C., is 18·5 per cent. (this is for twenty expansions; *vide* Fig. 18), while the corresponding overall efficiency for the steam-and-air turbine for the same initial temperature is 26 per cent. (*vide* Fig. 34). With reheating the overall efficiency of the single-fluid turbine becomes 25 per cent. This means that for the single-fluid turbine there is increase of 19·6 per cent. in the fuel economy by this reheating; in the case of the steam-and-gas turbine an increase of 21 per cent. Its application to the single-fluid turbine is, of course, abortive, on account of the lowering of the initial temperature and consequently low efficiencies.

The method by which the introduction of the additional

heat is effected is simply that of injection of the combustion mixture into the exhaust gases from the first turbine wheel. The system consists, essentially, of two expansion wheels, a boiler, a reheater, a regenerator and two pumps.

A diagrammatic representation of the system is shown in Fig. 45. The fuel is compressed into the furnace of a boiler at the same pressure as the steam in the boiler ; the steam space and the flue communicate, and the mixed products of combustion and steam are led into the high-pressure turbine wheel. The exhaust from this is led into the reheater, where it mixes with an additional charge of fuel at the atmospheric pressure. The gases then pass into the low-pressure turbine wheel, are expanded down to the terminal pressure of the cycle, and then exhausted through the regenerator. Steam is raised in the regenerator at the same pressure as that in the boiler. The compression pump, required to compress the fluid into the furnace, the exhausting pump, to remove the waste gases from the system, and the condenser, to take away the remaining heat in the working fluid, are not shown on the diagram.

IV. MIXED FLUID TURBINES USING OTHER CONDENSIBLE FLUIDS THAN STEAM.

Before leaving the question of the mixed-fluid turbine, there still remain to be considered turbines of this type employing accessory condensible fluids other than steam.

It must be confessed the ground here is very barren of any useful results ; it must, however, be summarily examined, if only to weed out and dismiss some cumbrous and unnecessary growths. The introduction of other liquids than steam into gas turbine systems takes the form of employing the waste heat of the exhaust gases to evaporate the liquids in question and extract power from them by expansion in a suitable turbine wheel. The four substances proposed, other than steam, by which regeneration may be effected are, carbon dioxide, ammonia, sulphur dioxide and ether. Of these four the first two may straightway be put out of court on account of the high pressures involved (at the temperature of the boiling-point of water the vapour pressure of ammonia is over 200 lbs. per square inch, and the pressure of carbon dioxide over

1,000 lbs., while the temperature of the exhaust from the gas turbine is 500° C.), the last on account of its excessive cost, if on no other count. The use of sulphur dioxide as a working fluid may be put out of court on the grounds of its corroding influence, for with the least admixture of water sulphuric acid is formed, and the total abolition of water is almost impossible.

Even were this not the case, however, it is not, from a purely thermodynamic standpoint, superior to water.

CHAPTER V. THE "EXPLOSION" GAS TURBINE

Of the three possible gas turbine cycles mentioned in Chapter I. only one cycle has so far been considered, namely, that type of gas turbine—mixed-fluid or otherwise—in which the heat is taken into the working fluid under the condition of constant pressure. The two other conditions of heat absorption which are possible are : (1) heat taken in at constant volume ; (2) heat taken in at neither constant volume nor constant pressure, but during some thermodynamic state between these two.

The process of introducing heat at constant volume into a gas by any process of external heating through metal walls is conceivable but plainly inadmissible in practice. Such need not be considered. The introduction of heat has, then, to be brought about by internal application. That is to say, the combustion mixture has to be ignited in a chamber (closed or otherwise). For this reason it has seemed best to class this type of turbine under the title of "explosion" gas turbines.

The two cycles in the explosion turbine mentioned above take place according as the turbine has the explosion chamber "closed" or "open." In the first case, the explosion occurs behind a valve, and the products of combustion are not admitted on to the turbine wheel until the explosion has ended and the maximum pressure has been reached. In the latter case the explosion takes place behind a column of still air, in an open chamber or a long tube, the back end of which is closed and the front end open on to the turbine wheel. In this case the pressure and volume of the gas actually change while the heat is being introduced into the fluid.

The most striking departure in turbines of the explosion type from those of the constant-pressure type is that the pressure drop upon the turbine wheel falls from a maximum value to zero, that the velocity on the blades varies, therefore, between maximum and minimum values, and the impulse on the turbine, instead of being continuous, is periodic. The nature of this periodicity, and the losses consequent upon it, will be considered later.

It is convenient, as well as more logical, to consider the two cycles of the explosion turbine separately. The closed chamber heat absorption at constant-volume type will be considered first, as it is more efficient, and has been more investigated than the open-chamber type. Though, from a purely chronological point of view, the latter has precedence.

THE CLOSED-CHAMBER EXPLOSION TURBINE.

Heat Taken in at Constant Volume.

A mixture of combustible gas and its requisite proportion of air is ignited in a chamber the exit from which is closed by means of a valve control. When the liberation of heat consequent upon the chemical combination has raised the fluid to a maximum pressure, the valve is opened, and the products of combustion allowed to expand upon the turbine wheel, down to the atmospheric pressure—or to any arbitrary terminal pressure adopted. The remaining heat from the exhaust gases is removed by a condenser or heat sump (regenerative or otherwise).

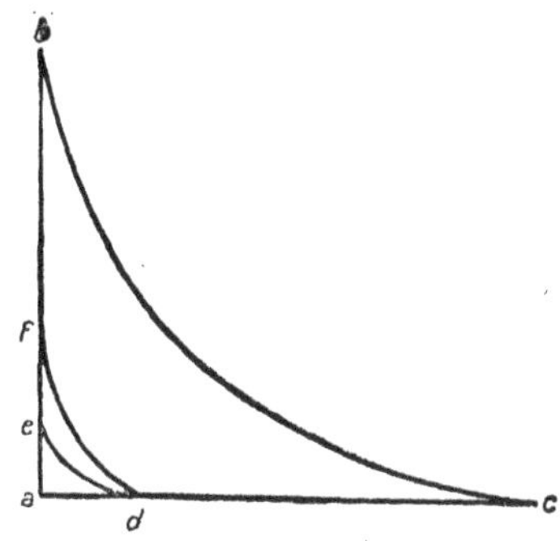

FIG. 46.—"PV" Diagram for Cycle II.

The gases may undergo initial compression, previous to ignition, or there may be no previous compression, as in the old Lenoir gas engine. The cycles are shown on the "PV" diagram in Fig. 46.

(*a*) *No Initial Compression.*—Heat is taken in at constant volume along *ab*; adiabatic expansion occurs along *bc*; heat is abstracted at *constant pressure* along *ca*.

(*b*) *With Initial Compression.*—The gases are compressed isothermally from *d* to *e* (or adiabatically from *d* to *f*); heat is introduced at constant volume from f to b ; adiabatic expansion occurs from b to c, and rejection of heat at *constant pressure* along cd. It will be noticed that these represent the ordinary gas engine cycles—Lenoir or Otto—with the following important variations : (1) the rejection of heat at constant pressure—*not* at constant volume ; (2) initial compression is isothermal instead of being adiabatic. The reasons for the adoption of the isothermal rather than the adiabatic compression applies here as in the case of the constant-pressure gas turbine ; they need not be reiterated here ; the reader is referred to pp. 17, 18. The point has not the instancy here that it has with the constant-pressure turbine.

The same cycles are shown in Fig. 47 on the entropy-temperature diagram. Absorption of heat at constant volume takes place along ab ; adiabatic expansion along bc ; rejection of heat at constant-pressure along ca. If initial compression takes place, the expansion is increased to bc' ; the rejection of heat at constant-pressure occurs along $c'd$, and the isothermal compression along da.

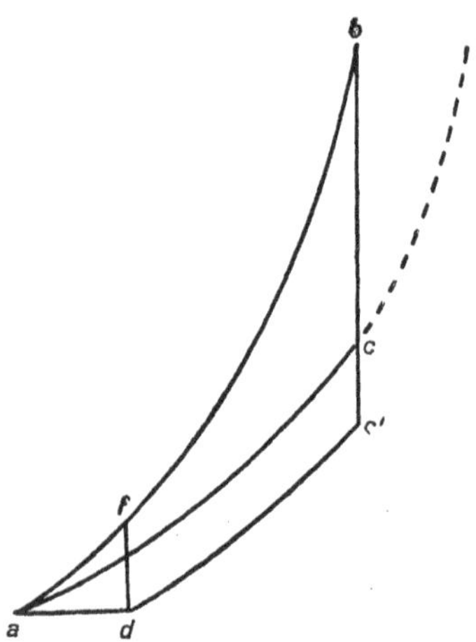

FIG. 47.—" ΘΦ " Diagram for Cycle II.

It will be seen that, in the case of no initial compression, the work done by the cycle is represented by the space lying between the curves for heat-absorption at constant volume and heat absorption at constant pressure; the boundary line to the right of the diagram being made by a vertical, whose position is determined by the initial temperature in the combustion chamber. This is, of course, on the purely theoretical assumption that there is no loss due to " contraction " (*vide infra*) or to any extraneous heat flow.

Before consideration of the explosion gas turbine as a heat engine it may be as well to devote a few paragraphs to a brief survey of the conditions of explosion in closed vessels.

EXPLOSION IN CLOSED VESSELS.

Unit weight of hydrogen gas gives, upon combustion with oxygen, 34,170 T.U. Now the combustion of 1 lb. of hydrogen produces 9 lbs. of water. If we take the specific heat of steam at constant pressure to be ·48, and to be constant over the temperature range involved, we get a theoretical temperature for the combustion of oxygen and hydrogen, under a constant pressure, of—

$$T = \frac{34,170}{\cdot 48 \times 9} = 7,910^\circ \text{ C.}$$

The latent heat of the formation of steam has, however, to be taken into account, therefore the true theoretical temperature of combustion becomes—

$$T = \frac{34,170 - 539}{\cdot 48 \times 9} = 7,785^\circ \text{ C.}$$

If we assume the combustion to have taken place in a closed chamber, we must substitute the specific heat at constant volume for the specific heat at constant pressure, viz., the value ·37 for ·48. This gives a theoretical temperature of explosion of 10,120° C. With the combustion of carbon in oxygen the temperature is even higher, reaching, under conditions of constant pressure, a temperature of 10,174° C. Thus the combustion of a mixture of oxygen and hydrogen in a closed chamber, in the proportion of their combining weights, should produce a pressure of 10,393/273 or 38·1 atmospheres, or a pressure of 560 lbs. per square inch.

In actual fact, neither the temperatures nor pressures calculated above are produced. The failure to attain to these temperatures is due to the phenomenon known as *dissociation*.

DISSOCIATION.

It has been asserted by Ostwaldt that *every* chemical action, no matter what its nature, is really a "reversible" reaction, and that the direction in which the action proceeds depends only upon the conditions of pressure and temperature under which the action is taking place, and that the apparent disavowal of this thesis in particular cases is but due to the fact that the range of physical conditions has not been sufficiently extended. In the case of combustion—or, to confine it to a particular case, with the combination of oxygen and hydrogen—the action is reversible well within the observable temperature range. The equation—

$$2H_2 + O_2 = 2H_2O$$

must be written in the form—

$$2H_2 + O_2 \rightleftarrows 2H_2O$$

the direction in which the chemical reaction occurs depending upon the temperature.

Reaction between oxygen and hydrogen will take place at even so low a temperature as 300° C., though several days are required to produce a small quantity of water. At 518° C. the union is complete after several hours. At 600° C. the action is rapid, but not explosive. At 700° C. the combination is almost instantaneous. Thus the action continues in this direction up to about 960 to 1,000° C. At this temperature the reversibility of the action appears (Deville); in other words, "dissociation" commences, and as the temperature rises above this value, the action begins to take place in the other direction, the water formed dissociating into its constituent elements. The action, which was before *exothermic*, now becomes *endothermic*, and a balance is soon reached. A limit is thus set to the maximum temperature possible upon combustion considerably lower than the calculated value. Deville's experiments on the temperature of the oxy-hydrogen flame gave a value of 2,800° C. From the experiments of Bunsen the highest maximum temperature that can be assumed is 3,800° C.; from the experiments of Mallard and Le Chatelier, 3,500° C.

THE COEFFICIENT OF CONTRACTION.

There is yet another matter which affects the possible pressure upon explosion, and that is the contraction in volume produced by the chemical reaction. In the case of the combustion of oxygen and hydrogen, for example, we have—

$$2H_2 + O_2 = 2H_2O$$
$$2 \text{ vols.} + 1 \text{ vol.} = 2 \text{ vols.}$$

Thus the volume of the gases after explosion is one-third less than the volume of the gases before explosion. This, of course, produces a corresponding lowering of the pressure from the value calculated from the temperature. It will be noticed that, where the gases are considerably diluted, this "coefficient of contraction" becomes of less account. Thus, if the combustible gases occupy one-fifth of the total charge, the contraction coefficient is reduced from ·333 to ·066.

TEMPERATURE OF EXPLOSION : THE IDEAL AND THE REAL.

If a mass of a perfect gas is exploded in a closed vessel, and T be the temperature before ignition, and T′ the temperature after ignition (° Absolute), and p be the absolute pressure before and p' the absolute pressure after combustion, we have, from Charles' Law—

$$T/T' = p/p'.$$

This does not take into account the effect of contraction.

If the coefficient of contraction is k, we have—

$$T/T' = (p/p') \, . \, k.$$

In this case, however, the true temperature cannot be calculated, as it is unknown how far dissociation has occurred, and consequently the amount of contraction is also unknown. It can only definitely be stated that the true temperature must lie between two values : (1) on the assumption that no contraction has taken place ; (2) on the assumption of complete contraction. Let these two temperatures be denoted, respectively, by the symbols θ and θ'.

	θ	θ'
2 vols. H + 1 vol. O (9·9 atmos. abs.)	2,449° C.	3,809° C.
2 vols. CO + 1 vol. O (10·8 atmos. abs.)	2,612° C.	4,140° C.

The lower temperature necessitates complete dissociation, the higher no dissociation. The actual temperature must lie between these two figures.

This matter of the temperature of explosion can best be dealt with by taking an actual example under conditions pertinent to the gas turbine. Consider a sample of Mond gas of the following composition :—

	By volume.
Hydrogen, H_2	20·5
Methane, CH_4	2·3
Carbon monoxide, CO . . .	15·7
Carbon dioxide, CO_2 . . .	11·5
Nitrogen, N_2	49·8
	100·0
Air (necessary for combustion) .	108·0
Total volume of mixture .	208·0

The following reactions take place :—

	Contraction.
$2H_2 + O_2 = 2H_2O$	1/3
2 *v*. 1 *v*. 2 *v*.	
$CH_4 + 2O_2 = CO_2 + 2H_2O$	0
1 *v*. 2 *v*. 1 *v*. 2 *v*.	
$2CO + O_2 = 2CO_2$	1/3
2 *v*. 1 *v*. 2 *v*.	

Contraction, therefore, occurs with 54·3 volumes out of the 208 volumes of the explosive mixture. The coefficient of contraction, then, for the whole charge, is—

$$k = \frac{54{\cdot}3}{3 \times 208} = {\cdot}087.$$

The thermal value of the combustion mixture is 550 T.U. per lb. The theoretical temperature of the fluid after combustion is, therefore—

$$\theta = 550/C_v = 3{,}060° \text{ C.}$$

taking specific heat at constant volume for the combustion mixture to be ·18. As the thermal value is here reckoned from the atmospheric temperature, the theoretical temperature of the fluid is 3,075° C.

Such a temperature is too high to be used in practice. Let 1,200° C. be adopted as a value for θ.

This lowering of temperature must be brought about by dilution with air; to obtain quantity of air added—

$$\theta = \frac{H}{C_v + x . C'_v},$$

where C_v is the specific heat at constant volume for the products of combustion, and C'_v that for pure air, and x the weight of air added.

Therefore, we have—

$$1{,}185 = \frac{550}{\cdot 18 + x . \cdot 172}.$$

This gives a value for x of ·67 lb.

Therefore, contraction on total mass of working fluid is—

$$k_T = \cdot 087/1\cdot 67 = \cdot 052.$$

Thus the temperature calculated from pressure $\theta = \cdot 948\, \theta'$, the actual theoretical temperature.

In the actual experiments made in respect to the temperatures and pressures of explosions in closed chambers the temperatures calculated from the observed pressures differed very widely from the calculated values. In the classic experiments carried out by Mr. Dugald Clerk the observed temperature fell to a value of about half the theoretical temperature. This can only be accounted for in the following ways:—

(1) Loss of heat due to radiation and conduction from the explosion chamber.
(2) Incomplete combination due to dissociation effects.
(3) Increase in specific heat with rise in temperature.
(4) Divergence from the gas laws; *i.e.*, Charles' Law, Boyle's Law, Van der Waals' equation, etc.

With the charge so diluted as not to allow of a theoretical temperature exceeding 1,200° C. to 1,400° C., the last three items should not be of pressing importance. The large insufficiency noted by Mr. Dugald Clerk, even with diluted mixtures, and correspondingly low temperatures must be considered to arise mainly from the conduction and radiation of heat from the explosion chamber; this factor would also be of more moment in such cases where the rate of combustion is relatively slow. This loss is, then, of an arbitary nature and controllable by insulation, or palliatable by regenerative processes. In the following consideration of the explosion gas turbine as a heat engine the theoretical temperatures and pressures will be assumed as a basis of calculation, correction being made for dissent from theoretical conditions by the introduction into the calculations of a factor, assumed at a certain value, and to a large extent variable by the control of peculiar conditions.

The author feels that, unsatisfactory as such a procedure is, still it is the only sane and logical way by which a concrete concept of the explosion turbine as a heat engine can be achieved. The explosion turbine, of all others, presents the greatest difficulty to theoretical analysis, and involves a number of assumptions that only actual experiment, and what is more, an extended series of experiment, can adequately support.

THE EXPLOSION TURBINE AS A HEAT ENGINE.

(A) No Initial Compression (Fig. 48).

A mass of combustible mixture is contained in a closed vessel under conditions, T_0, p_2, v_1, The mass is then ignited, v_1 remains constant, p_2 increases to p_1, T_0 to T_1. Heat is taken in, therefore, during this stage, at constant volume. Also we have, $p_1/p_2 = T_1/T_0$, the temperature being measured on the absolute scale. The gas is then allowed to expand from p_1v_1 to p_2v_2. Heat is rejected at constant volume from v_2 to v_1. The work done during the expansion is only that portion of the "PV" diagram lying beneath the expansion curve, minus the area lying below the line of terminal pressure. This is represented by the equation—

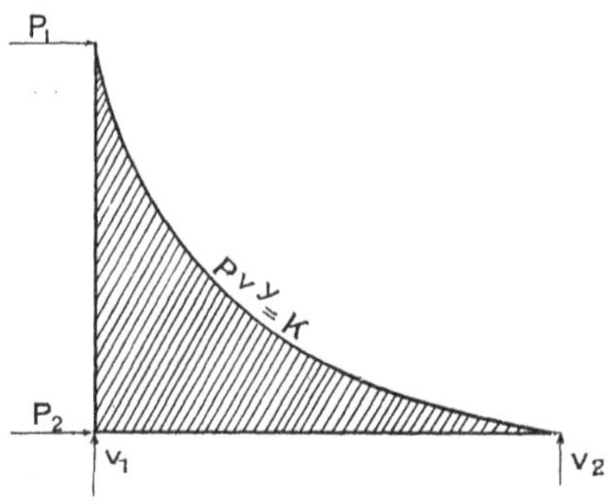

FIG. 48.—"PV" Diagram for Explosion Turbine. (No initial compression.)

$$W = \int_{v_1}^{v_2} p \, . \, dv - p_2 (v_2 - v_1).$$

Therefore we have—

$$W = \frac{p_0 v_0 . R}{\gamma - 1}(1 - 1/R^{\frac{\gamma-1}{\gamma}}) - p_0 v_0 (R^{\frac{1}{\gamma}} - 1)$$

where p_0 and v_0 are the pressure and volume of 1 lb. of the gas at atmospheric temperature and pressure, and R is the ratio $p_1/p_2 = T_1/T$.

The heat taken into the system is—

$$H = C_v'(T_1 - T_0)$$

Therefore thermal efficiency—

$$\eta = W/H.$$

If regeneration is employed, the heat from the exhaust gases goes to heat the supply to the explosion chamber; then, if T' is the temperature at which the exhaust gases leave the regenerator, the heat lost is—

$$h = C_p(T' - T_0)$$

and the thermal efficiency—

$$\eta = \frac{W}{W + h}.$$

In a particular case let $T_1 = 1{,}440°$ A. Then $R = 5$. This gives a value for W of 46·2 T.U.

The heat put in, H, equals 207 T.U. ($C_v = {\cdot}18$). Therefore, without regeneration—

$$\eta = 46{\cdot}2/207 = 22{\cdot}3 \text{ per cent.}$$

With regeneration—

$$\eta = \frac{46{\cdot}2}{46{\cdot}2 + 50},$$

taking $T' = 215°$ C. (*vide supra*). Therefore—

$$\eta = 48 \text{ per cent.}$$

If the thermodynamic efficiency of the turbine be 70 per cent., then we get a value for the overall efficiency of 33·6 per cent.

It must be borne in mind, however, that the temperature and pressures obtained by gaseous explosion in closed vessels always fall far short of the theoretical values; some factor giving the ratio between the real and the ideal must be chosen by which the above-obtained efficiency may be multiplied. With the series of experiments conducted by Mr. Dugald Clerk this factor was about ·5. With precautions against the loss of heat by conduction and radiation this value might be increased. Let this factor be taken, then, at 70 per cent. This gives a final overall efficiency of 23·5 per cent.

Comparing this to the overall efficiency for the steam-and-gas turbine working with the same initial temperature, it will be seen that there is here a decrease in efficiency of nearly 20 per cent.

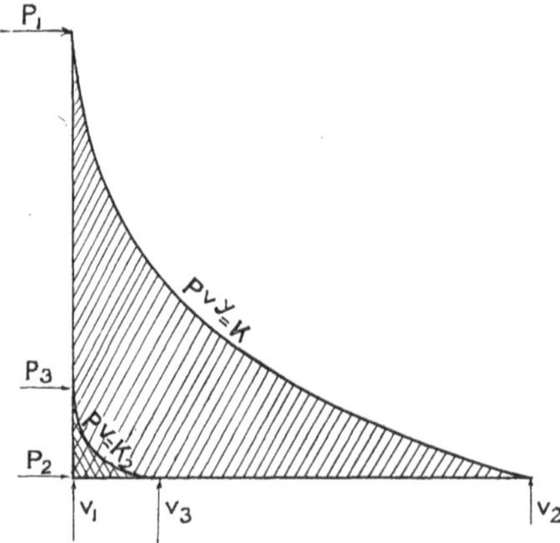

FIG. 49.—"PV" Diagram for Explosion Turbine. (With initial compression.)

(B) With Initial Compression (Fig. 49).

A mass of combustible fluid is compressed isothermally from a condition $p_2v_3T_0$ to one of $p_3v_1T_0$. Heat is then introduced at constant volume until the condition $p_1v_1T_1$ is reached. Adiabatic expansion takes place from p_1v_1 to p_2v_2, the temperature falling to value T_2. Heat is abstracted at constant pressure from T_2 to T.

The work done equals, as before—

$$W = \frac{p_0v_0' \cdot R'}{\gamma - 1}\left\{1 - \frac{1}{R'^{\frac{\gamma-1}{\gamma}}}\right\} - p_0v_0'\left(R'^{\frac{1}{\gamma}} - 1\right)$$

save that R′ now equals R.r, where $R = T_1/T$ and r is the compression ratio in the pump. Also v_0' equals v_0/r ; p_0 is the atmospheric pressure.

The negative work done by the pump—

$$W_n = p_0v_0 \log_e . r - p_0v_0\left(\frac{r-1}{r}\right)$$

since the pump only has to compress the gases into the closed vessel until the maximum pressure is reached, and consequently the work done is only that represented by the portion of the "PV" diagram lying underneath the expansion curve minus the area below the terminal pressure line.

Assuming $r = 2$, and R, as in the former case, equals 5, this gives a value for the positive work done by the turbine of W = 76 T.U. and for the negative work done by the pump of 3·6 T.U. Nett work done by the cycle equals, then, 72·4 T.U. The heat put in equals, as in the case of no initial compression, 207 T.U. Therefore—

$$\eta = \frac{72{\cdot}4}{207} = 35 \text{ per cent.}$$

If regeneration takes place, we have—

$$\eta = \frac{72{\cdot}4}{72{\cdot}4 + 50} = 59 \text{ per cent.}$$

Suppose the pump to be driven by a steam turbine, the steam for which is raised by the waste heat from the turbine exhaust. With an expansion range of 150 lbs. to 1 lb. the thermal efficiency of the steam is 28·3 per cent. Taking the thermodynamic efficiencies of the steam turbine and the pump to be each 70 per cent. we have heat required for the pump—

$$h' = 3{\cdot}6/{\cdot}139 = 26 \text{ T.U.}$$

If the thermodynamic efficiency of the gas turbine be also taken at 70 per cent., then overall efficiency—

$$\text{E} = \frac{76 \times {\cdot}7}{76 + 50 + 26} = 35 \text{ per cent.}$$

or, taking the efficiency of the explosion chamber at ·7—

$$\text{E} = 24{\cdot}5 \text{ per cent.}$$

This again falls short of the overall efficiency given by the steam-and-gas turbine with the same initial temperature.

The above is a very tentative *examen* of the explosion turbine as a heat engine. The most favourable efficiencies obtainable have been ascertained, and it may be pointed out that even were the expansion ratio to be increased beyond the practical limit, but a small gain in efficiency could result.

The system suffers thermodynamically from the fact that only that portion of the " PV " diagram underlying the expansion curve is available for work. True, the heat put into the system is decreased by the ratio C_v/C_p, but the work done, as compared to the constant-pressure turbine, is decreased to an even larger extent.

The great and pregnant point in favour of the explosion turbine as compared to the constant-pressure turbine, mixed-fluid or otherwise, does not lie on the side of thermal efficiency. It lies in the elimination, or at least in the great diminution, of the negative work of the pump.

There are three great difficulties to be contended with in regard to the explosion turbine. The first two are theoretical in nature, and are concerned with heat losses inherent, though to some extent controllable, in the system. They are :—

(1) Heat losses from the combustion chamber.

(2) Energy losses due to the change of velocity on the wheel.

The third difficulty is of purely a practical nature, though perhaps the most potent of them all. This is the very doubtful possibility of getting a valve to work satisfactorily for any considerable period of time between the explosion chamber and the wheel (*vide* Chapter XI.).

The first of these losses, namely the heat flow from the combustion chamber, is only to be found empirically ; beyond the fact that this loss is proportional to the fourth power of the temperature difference between the inside and outside of the explosion chamber, and that it is directly proportional to the surface of the chamber, we have no sure theoretical means for estimating that loss. The second loss, due to the variable velocity of the fluid, can be computed (approximately) from theoretical considerations.

Neglecting the secondary loss, and taking the thermodynamic efficiency of the turbine to be still 70 per cent., the effect of the first loss on the efficiency of the explosion turbine can only be definitely arrived at by the consideration of actual experimental data. Fortunately here we are not without that data which is so conspicuously lacking in connection with other types of gas turbines. A gas explosion turbine has been built of some size (1,000 H.P.) and extensive experiments made with it by Herr Holzwarth of Mannheim ; the practical work of Herr

Holzwarth need not be considered here, nor the construction and details of his turbine gone into, as a later chapter (Chapter XI.) is concerned with that subject, and repetition would be useless and cumbersome. The experimental data for the working out of a particular case will alone be quoted here.

There is another source of experimental data bearing upon the heat losses in explosion chambers, and that is the classic records of Mr. Dugald Clerk. The efficiency of an explosion turbine will be worked out from the data taken from each of these sources ; the conditions of temperature and pressure are, as near as possible, chosen in agreement.

(A) THE EFFICIENCY OF THE EXPLOSION GAS TURBINE, BASED ON THE EXPERIMENTS OF MR. DUGALD CLERK.

In a series of experiments made by Mr. Clerk upon the explosion of gas and air in closed vessels it was found that a mixture composed of 1 volume of Oldham coal gas with 11 volumes of air, gave, upon ignition, a maximum pressure of 5·15 atmospheres absolute, corresponding to a temperature of 1,220° C.

The thermal value of the actual gas used is not given, but that for Glasgow coal gas is given, which is mentioned as being similar to the Oldham gas. The thermal value has, therefore, been approximated at 11,500 T.U. per lb., and the specific gravity at ·036 lb. per cubic foot.

This gives a ratio of gas to air by weight of 1 : 24·7. Therefore, heat put into 1 lb. of working fluid—

$$H = 11{,}500/25{\cdot}7 = 450 \text{ T.U.}$$

Let the initial compression before explosion be 1·52 atmospheres. This gives a value for the initial pressure of 7·82 atmospheres, and a value for R of 7·82. Using the equation given on p. 114, we have—

$$W = 1/m \left\{ \frac{p_0 v_0 \, . \, R}{\gamma - 1} \left(1 - \frac{1}{R^{\frac{\gamma-1}{\gamma}}} \right) - p_0 v_0 \left(R^{\frac{1}{\gamma}} - 1 \right) \right\}$$

m being equal to the ratio, $v_3/v_1 = 1{\cdot}52$.

Therefore $W = 66{\cdot}7$ T.U.

Writing off the negative work against the waste heat, we have—

$$\eta = W/H = 66{\cdot}7/450 = 13{\cdot}4 \text{ per cent.}$$

Taking the thermodynamic efficiency of the turbine at 70 per cent., the overall efficiency, E, equals 9·4 per cent.

If regeneration is taken into account, there are three heat losses to be considered :—

(1) Heat loss in regenerator exhaust, with $T' = 215°$ C., $h = 50$ T.U.

(2) Heat required to drive pump ; if steam driven as above, heat loss = 10·4 T.U.

(3) Heat loss due to inefficiency of combustion chamber.

The efficiency of the combustion chamber—

$$\eta_e = (T_1 - T_0)\, C_v/H \qquad T_1 = 1{,}220° \text{ C.} \quad T_0 = 17° \text{ C.}$$

$$= 49{\cdot}2 \text{ per cent.}$$

Therefore, heat loss from combustion chamber—

$$h' = {\cdot}518\, H = 233 \text{ T.U.}$$

Therefore, overall efficiency—

$$E = \frac{66{\cdot}7 \times {\cdot}7}{233 + 66{\cdot}7 + 50 + 10{\cdot}4}$$
$$= 13 \text{ per cent.}$$

(B) THE EFFICIENCY OF THE EXPLOSION GAS TURBINE, BASED UPON THE EXPERIMENTS OF HERR HANS HOLZWARTH.

The initial temperature of the charge, T_0, was 15° C.

The heat supplied to the explosive mixture was 228 calories per kilo ; the thermal value of the gas 1,179 cal./kilo.

The initial temperature, T_1, was 1,398° A., calculated from the equation—

$$p'_1 = \phi \cdot p \frac{T_1}{T_0}$$

where ϕ = coeff. of contn.
= 95·8 per cent.

The initial pressure, p_1, equalled 7·08 atmospheres.
The compression ratio for the pump, as before, 1·52.
Heat supplied, therefore, per lb. of working fluid—

$$H = 230 \text{ T.U.}$$

Thus the work done by the fluid on expansion—

$$W = 1/m \left\{ \frac{p_0 v_0 . R}{\gamma - 1} \left(1 - \frac{1}{R^{\frac{\gamma-1}{\gamma}}}\right) - p_0 v_0 \left(R^{\frac{1}{\gamma}} - 1\right) \right\}$$
$$= 56·8 \text{ T.U.}$$

Writing off negative work against waste heat, and assuming no regeneration, as in the case of the Holzwarth turbine—

$$\eta = W/H$$
$$= 56·8/230$$
$$= 24·7 \text{ per cent.}$$

With the thermodynamic efficiency of the turbine at 70 per cent. we have overall efficiency—

$$E = 17·3 \text{ per cent.}$$

Taking regeneration into account, we have a loss of 50 T.U. in exhaust from the regenerator, a loss of 10·4 T.U. on the steam-driven pump, and the loss due to the efficiency of the explosion chamber. In this case—

$$\eta_e = \frac{(1{,}398 - 288)}{230}$$
$$= 87 \text{ per cent.}$$

Therefore, heat loss from combustion chamber—

$$h' = ·13 \text{ H} = 30 \text{ T.U.}$$

Therefore, overall efficiency—

$$E = \frac{56·8 \times ·7}{56·8 + 10·4 + 50 + 30}$$
$$= 27 \text{ per cent.}$$

It will be noticed that there is a very considerable divergence between the "efficiency of explosion" in the experiments of Mr. Dugald Clerk, and those of Herr Holzwarth. The best explosion efficiency obtained by Mr. Clerk appeared to be—

$$\theta'/\theta = 63 \text{ per cent.}$$

(in case taken, 49·2 per cent.), compared to the efficiency obtained by Herr Holzwarth of—

$$\theta'/\theta = 87 \text{ per cent.}$$

The factor in question depends primarily on conditions of temperature and pressure, and upon the construction of the explosion chamber.

Although the limits of overall efficiency in the case of the explosion turbine do not admit of any ready formulation (owing to the variability of heat losses from the explosion chamber) yet the limits of the thermal efficiency of the cycle are very definite.

As in the case of the constant-pressure turbine, let the temperature upon the turbine blades, T_2, be 500° C. If there is no initial compression, the maximum temperature which it is possible to have in the explosion chamber is—

$$\theta_1 = \theta_2 \, . \, R^{\frac{\gamma - 1}{\gamma}}$$

where $$R = \left(\frac{\theta_1}{\theta_0}\right).$$

If, however, there is initial compression, then the maximum temperature is only limited by the velocity that the turbine wheel can absorb, and—

$$\theta_1 = \theta_2 \, . \, (R \, . \, r)^{\frac{\gamma - 1}{\gamma}}$$

where r is the ratio of compression in the pump. The heat put in being—

$$H = C_v \, . \, (\theta_1 - \theta_0.)$$

The thermal efficiencies possible with varying temperatures are shown clearly on the temperature-entropy chart, Fig. 50. With no initial compression the work done is represented by the area bounded by the curve of heat absorption at constant volume and the curve of heat absorption at constant pressure, and the vertical of adiabatic expansion. The maximum work available in this manner is shown by the shaded area, the expansion being from $\theta_1 = 1{,}210°$ A. to $\theta_2 = 773°$ A., or an expansion of 4·2. The work done by the turbine, the work done by the pump, and the heat put in to the system are further shown for values of θ_1 from 1,300° A. to 2,000° A. at intervals of 100°.

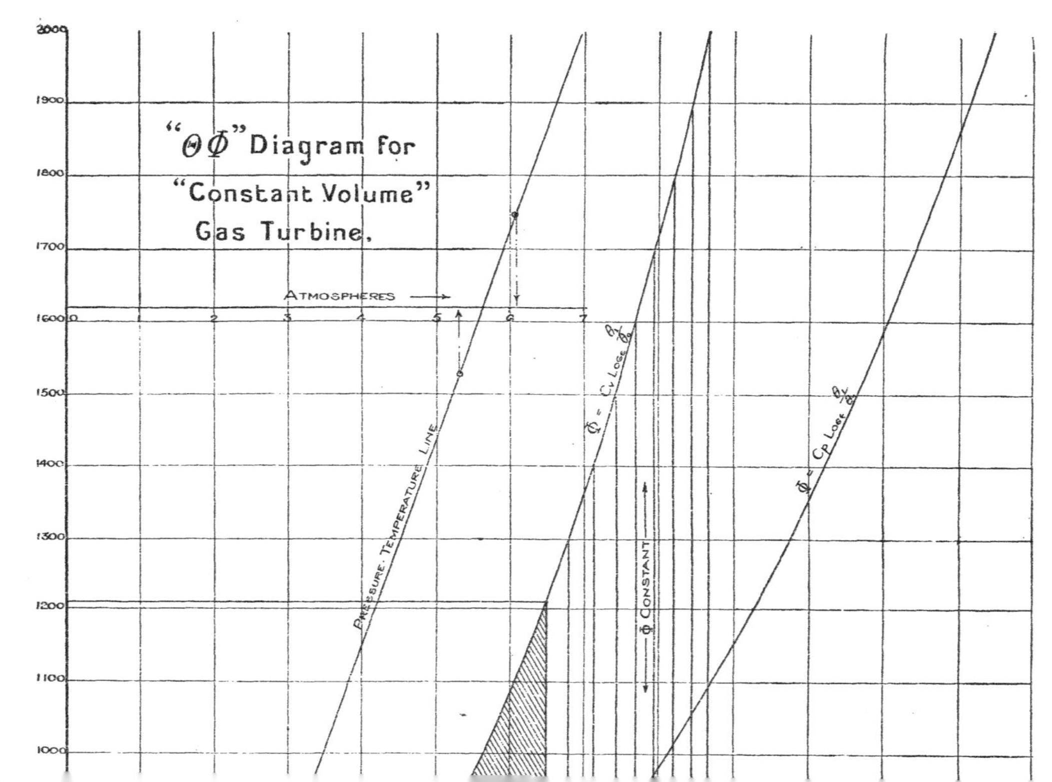

"ΘΦ" Diagram for
"Constant Volume"
Gas Turbine.
Atmospheres
0
1
2
3
4
5
6
7
Pressure · Temperature Line
Φ = Cv Loge θ/θ0
Φ = Cp Loge θ/θ0
Φ Constant
2000
1900
1800
1700
1600
1500
1400
1300
1200
1100
1000
θ1 Max. N = Cpn
972°C
ATURE °C-abs

θ_2. Max Cnst.
500°C
θ_0. Atmos. T
15°C
800
773
700
600
500
400
300
288
273
200
100
0
Line of Max. Temp. on Blades
Work Done No Comp.
Line of Atmos. Temp.
Work on Pump.
Heat Loss from Regr.
0
·1
·2
·3
·4
·5
Entropy

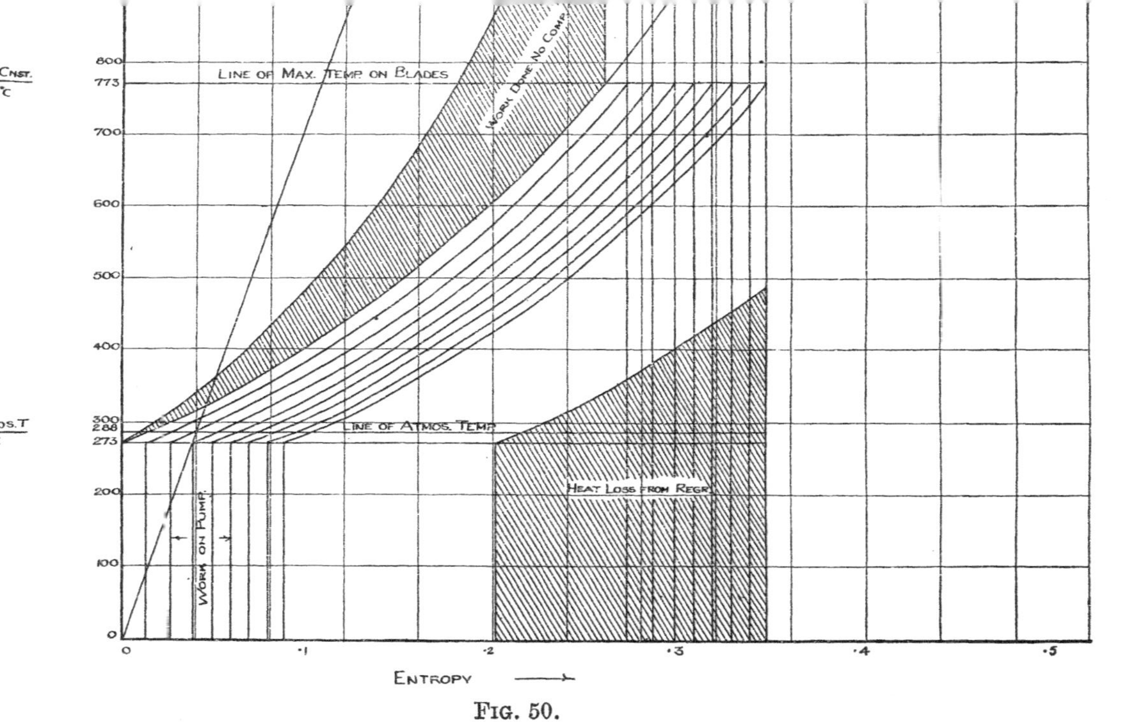

FIG. 50.

The temperature-pressure line, it must be noted, only refers to cases of no initial compression.

In the case of regeneration, the only heat theoretically put into the system is the waste heat from the regenerator together with the work done by turbine and pump. The former item (assuming a constant temperature at exit of regenerator) is a constant value. It is shown by the shaded area in the right-hand corner of the chart. In this instance the temperature of the exhaust from the regenerator is taken at 215° C. The " $\theta\phi$ " chart is drawn on the assumption that at 0° C. $\phi = 0$.

It was mentioned above that one of the chief objections to the closed-chamber explosion turbine was the presence of a valve between the explosion chamber and the turbine wheel. This question, together with other questions of detail and construction, will be treated of later in the discussion of the experimental work of Herr Holzwarth.

LOSS DUE TO VARIABLE VELOCITY.

We have to consider the fact that the velocity of the gases flowing from the combustion chamber on to the wheel fall in velocity from a maximum to a minimum, or may be a zero, value, while the turbine wheel itself must, of necessity, run at a fixed peripheral speed. It is now to be investigated what the loss is, due to this periodicity of impulse.

The conditions affecting the velocity of the flow are as follows : The gases are contained in a closed chamber behind a valve at a maximum pressure, p_1. The valve is then opened. The first particle that leaves the chamber does so under the influence of an expansion ratio p_1/p_0, which is the maximum expansion ratio. The first element of gas has, therefore, the maximum velocity. The pressure in the chamber has, however, now been reduced to a pressure $(p_1 - \delta p)$, and the expansion ratio for the next issuing element of the gas is $(p_1 - \delta p)/p_0$ or $R - \delta R$. The velocity of this element is less than that of the preceding one. Thus the velocity of the gas-flow falls until the pressure in the chamber equals the pressure of the surrounding medium. The ratio of expansion now becomes unity, and the last element of air leaves the combustion chamber with no velocity ; that is to say, it does not leave the chamber at all.

In point of fact, this is not actually the case. If the gaseous flow was allowed to take place infinitely slowly, such a condition of affairs would be brought about. But with sudden release of the valve, as is the case in practice, the preceding element of gas has a certain momentum, which acts in the nature of a "pull" upon the element of gas immediately behind it. Thus the final velocity is never zero, but some positive quantity; the result being that a momentary drop in pressure is produced in the explosion chamber to some value below the pressure of the surrounding medium.

Consider 1 lb. of fluid that has been exploded behind a valve, and is therefore at a certain pressure above that of the atmosphere outside the vessel. Let the valve now be opened. The initial portion of fluid, directly behind the valve, immediately acquires the maximum velocity due to the initial pressure difference. As the fluid leaves the chamber, however, the pressure falls, and the velocity of the fluid diminishes until a minimum, or maybe a zero, value is reached. If the initial pressure difference had been maintained throughout the efflux of the gases, the fluid would have remained at its maximum veloicty, V_1, and the energy of the moving mass would have been $V^2_1/2g$. If the velocity of the fluid falls to a final value V_2, and the velocity gradient is in the form of a straight line, the energy given out by 1 lb. of gas is—

$$E = \frac{1}{2g}\int_0^1 V^2 dm$$

where dm represents any small element of mass in the effluent gases.

Since $V = m \tan\theta + V_2$, where $\tan\theta$ is the slope of the velocity mass curve $= (V_1 - V_2)/1$,—

$$E = \frac{1}{2g}\int_0^1 (m(V_1 - V_2) + V_2)^2 dm$$

$$= \frac{1}{6g}\left\{V^2_1 + V_2^2 + V_1V_2\right\}.$$

Let $V_2 - kV_1$, where k is a fraction less than one. Then—

$$E = \frac{V^2_1}{2g} \cdot \frac{1}{3}(k^2 + k + 1).$$

Thus, the actual energy equals the maximum energy possible,

if the initial pressure difference remained constant, multiplied by $\frac{k^2 + k + 1}{3}$ — where k equals the fraction that the final velocity is of the initial velocity. When the final velocity value is zero—*i.e.*, when $k = 0$—the energy available becomes one-third of that due to the initial pressure drop.

Fig. 51 is a mass-velocity and mass-velocity squared diagram

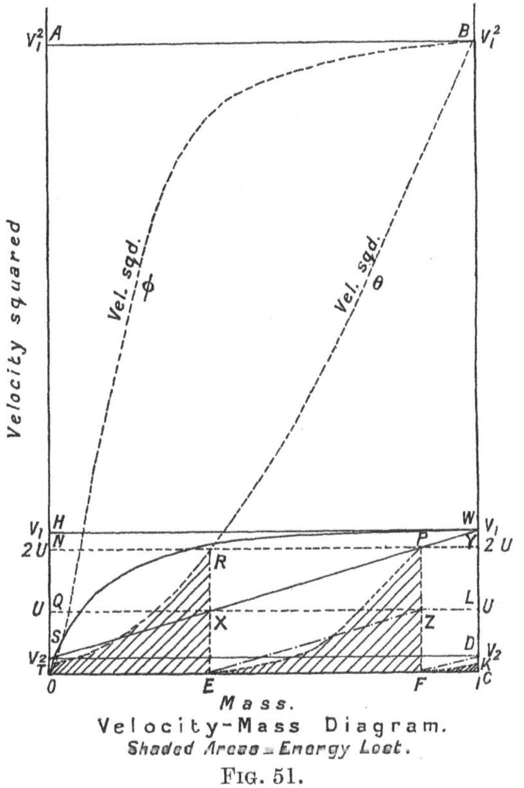

FIG. 51.

showing the loss in energy due to velocity variations. Here the velocity is assumed to fall from V_1 to V_2 along the straight line V_1XV_2, and the energy is represented by the area beneath the curve $B\theta T$. u is the peripheral velocity of the turbine wheel. The shaded areas represent the energy lost owing to the fixed velocity of the wheel and the varying velocity of the fluid. The area FKC represents energy lost while the fluid falls from a velocity V_1 to a value equal to $2u$. The area EPZF represents energy lost while the fluid falls from a velocity $2u$ to a value u, and the area TRXEO represents energy lost while the fluid falls from the velocity u to the final velocity V_2.

In the above velocity-mass diagram it is assumed that the velocity falls along a straight line. If the value of V_2 be taken at zero (a not possible case, but the final velocity must nearly approximate to this value), we have, total energy of effluent mass—

$$F = 1/3 \frac{V^2{}_1}{2g}.$$

Let the peripheral velocity of the wheel, u, be equal to $V_1/3$. Then the total energy lost due to the variation in velocity of the fluid will be—

$$f = 3 \int_0^1 \frac{u^2}{2g} \cdot \delta m.$$

But $u = V_1/3$, therefore—

$$f = 1/27 \cdot \frac{V^2{}_1}{2g}.$$

This gives a value for the percentage loss of—

$$f/F = 1/9 = 11{\cdot}1 \text{ per cent.}$$

This, however, only gives us the loss under conditions in which the velocity-mass curve is a straight line. This is a purely arbitrary assumption. It is possible, however, to construct the actual velocity-mass curve, or, more properly speaking, to plot the change in velocity with the fall in pressure.

Fig. 52 shows the velocity thus plotted for ten expansions on the "PV" diagram; the expansion being adiabatic, pv^γ = constant. The curve AB represents the adiabatic expansion. The curve FD represents the change in energy with fall in pressure, or a curve whose slope is—

$$\frac{\delta\,(\int p\delta v)}{\delta v}.$$

The area FED equals, of course, the area AHB.

The pressure, velocity, and energy are measured on the

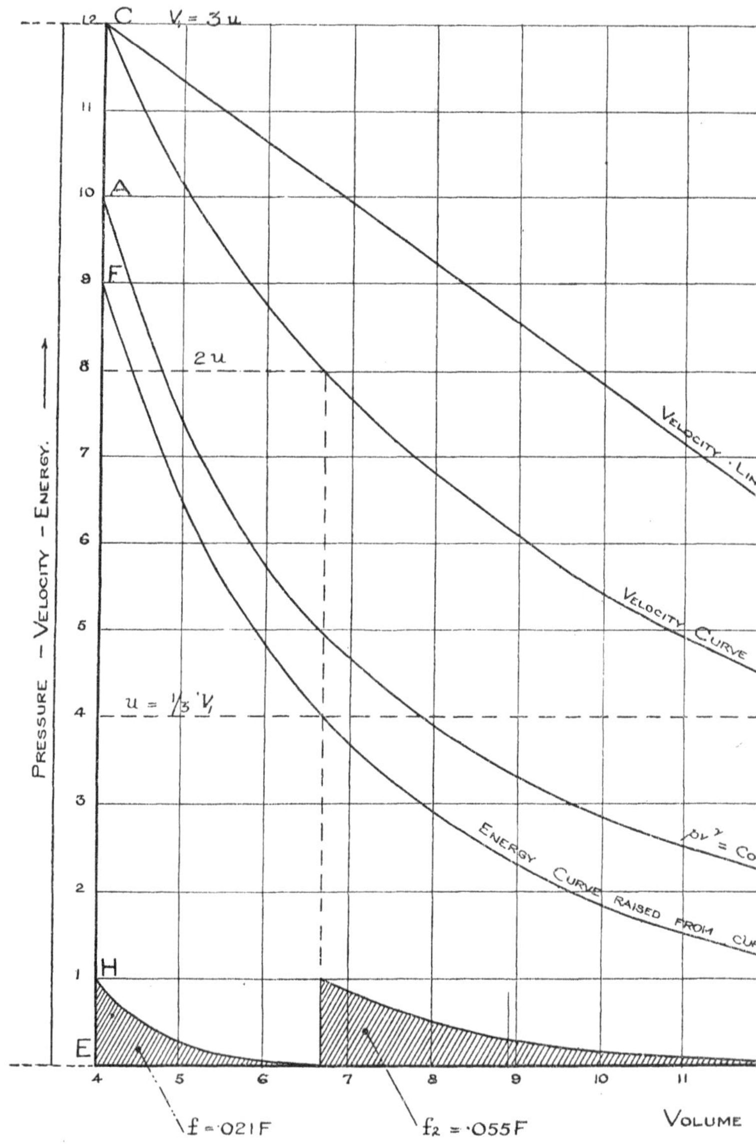

C
$V_1 = 3u$
A
F
2u
$u = \frac{1}{3} V_1$
H
E
Pressure – Velocity – Energy.
12
11
10
9
8
7
6
5
4
3
2
1
4
5
6
7
8
9
10
11
Velocity Lin
Velocity Curve
$pv^{\gamma} = Co$
Energy Curve raised from cur
$f = \cdot 021F$
$f_2 = \cdot 055F$
Volume

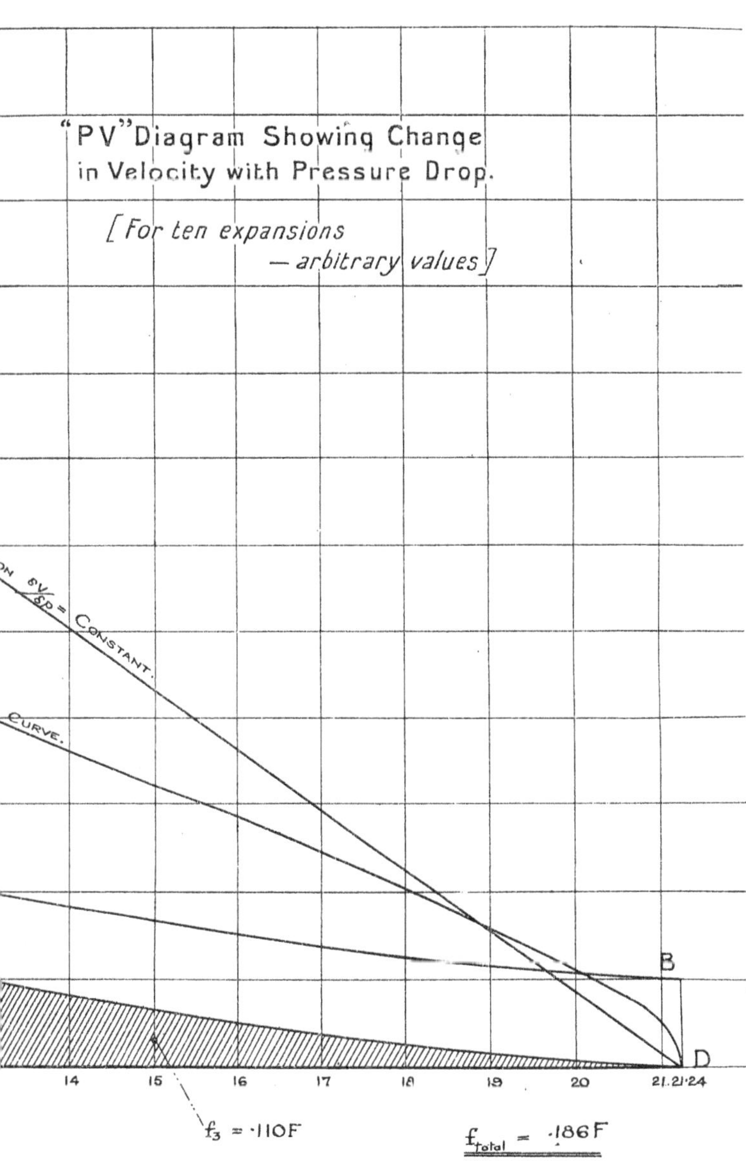

"PV" Diagram Showing Change
in Velocity with Pressure Drop.
[For ten expansions
— arbitrary values]
δV/δp = Constant.
Curve.
B
D
14
15
16
17
18
19
20
21.21·24
f3 = ·110F
ftotal = ·186F

vertical and in the same units, there being no absolute value assigned to these quantities, all that is needed being the percentage loss of energy to velocity variation.

It is assumed that the peripheral velocity, u, is a third of the maximum velocity, V_1. Parallels for u and $2u$ are drawn. The three losses of energy are represented by the three shaded areas, f_1, f_2, f_3 ; f_1 represents the loss in energy while the velocity falls from V_1 ($= 3u$) to value, $2u$; f_2 that lost in falling from $2u$ to u ; f_3 that lost in falling from u to 0.

The values of these losses are—

$$\begin{aligned} f_1 &= \cdot 021 \text{ F} \\ f_2 &= \cdot 055 \text{ F} \\ f_3 &= \cdot 110 \text{ F} \\ \hline f_{\text{total}} &= \cdot 186 \text{ F}. \end{aligned}$$

If the velocity had fallen along a straight line (CQD in Fig. 52) the energy loss would have been, as above, ·111° F. Here it is ·186° F. This increase is due to the fact that the actual velocity curve lies below the straight line CQD. If the velocity had so varied as to produce a velocity curve lying *above* the line QCD (as in WRO and BϕO, Fig. 51), f/F would become smaller.

It is to be seen how this loss in energy affects the value chosen for the thermodynamic efficiency of the turbine in the preceding chapters. This value was taken at ·7 for the various efficiency determinations. This included all the losses enumerated on pp. 77, 78.

The loss due to residual velocity is, in the constant-pressure turbine, recoverable, all save that due to the kataxial component. This is the same both for the explosion turbine and the constant-pressure turbine. The other losses noted are common to both turbines. On a comparative basis, therefore, it is fair to assume a loss of 30 per cent. in the explosion turbine due to causes other than variability in velocity. The loss due to the variability of the velocity is 18·6 per cent., if the velocity drops to a zero value. In reality, the loss would be slightly less than this. The thermodynamic efficiency, therefore, of the explosion turbine will be 51·4 per cent. In respect to the fact that the value of the final velocity is more than zero, we may take the thermodynamic efficiency for the explosion turbine,—

$$E_x = 55 \text{ per cent.}$$

The experimental values for E_x obtained by Herr Holzwarth varied from about ·4 to ·58 ; this is well in agreement with the above estimated value.

We can now apply this value for the thermodynamic efficiency of the turbine to the calculation of the overall efficiency of the examples of gas turbines worked out on pp. 118, 119.

(A) Mr. Dugald Clerk's Data.

I. No regeneration :

$$\eta = 13{\cdot}4 \text{ per cent.}$$

Overall eff. $\eta \,.\, E_x = 7{\cdot}4$ per cent.

II. With regeneration.

Overall eff. $\eta \,.\, E_x = 10{\cdot}2$ per cent.

(B) Herr Holzwarth's Data.

I. No regeneration :

$$\eta = 24{\cdot}7 \text{ per cent.}$$

Overall eff. $\eta \,.\, E_x = 13{\cdot}6$ per cent.

II. With regeneration :

Overall eff. $\eta \,.\, E_x$ 21·6 per cent.

OPEN CHAMBER EXPLOSION TURBINE.

Heat Taken in at neither Constant Volume nor Constant Pressure, but under some Condition varying between these two States.

If a mass of combustible fluid is ignited in a chamber that has a completely free exit, the heat is added at constant pressure, the rise in temperature producing change in volume only. With an aperture open to the atmosphere, however, the air immediately in front of the aperture is still. This air has to be set in motion before the volume expansion from the chamber can take place. If a certain mass, M, of air be regarded as blocking the exit from the chamber, and if the velocity of volume increase be V, we have—

$$Ft = MV$$

where F is the force acting, and t the time during which it acts. If the area of the aperture be A, we have—

$$p = \frac{MV}{At}$$

where p equals the rise in pressure produced in the explosion chamber above the pressure of the medium without.

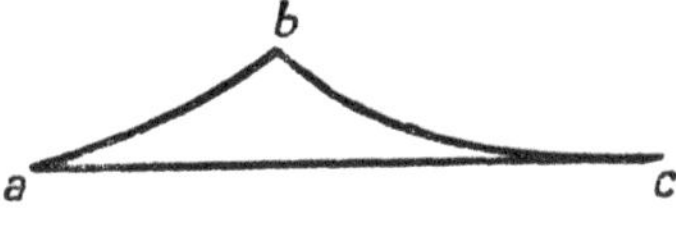

FIG. 53.—"PV" Diagram for Cycle III.

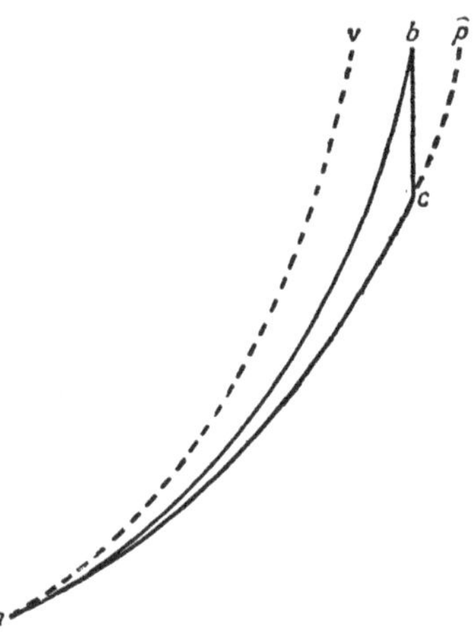

FIG. 54.—"$\theta\phi$" Diagram for Cycle III.

If the fluid be ignited in a chamber, large in capacity but narrow of exit, then the value of p is still further increased, the highest value being reached when the aperture vanishes and a condition of heat absorption at constant volume is attained. The value of the initial pressure may be brought about in a manner other than that of constricting the exit from the combustion chamber. The chamber may take the form of a long pipe, closed one end, and filled with still air. At the closed end of this pipe the charge is introduced and ignited. The sudden expansion sets the long column of air in motion, the inertia of which causes a rise in pressure of the gases forming the charge, a mean velocity of the total mass of fluid in the pipe being soon attained. It will be seen that in all of these instances the heat is put into the working fluid while both the pressure and volume change. The cycle is shown in Fig. 53 on the "PV" diagram. The pressure rises along ab at the same time as the volume is increasing; adiabatic expansion to the original pressure occurs along bc, and heat rejection at constant pressure along *ca*. The same cycle is shown on the " θ ϕ " diagram in Fig. 54. The curve of heat absorption lies between that of heat absorption at constant pressure, *ap*, and that at constant volume, *av*, and is represented by the curve, *ab*; adiabatic expansion takes place along *bc*, and rejection of heat at constant pressure along *ac*.

It is not possible to ascertain the theoretical thermal efficiencies of turbines of this type, for it is not possible to estimate otherwise than by actual experiment the pressure, temperature and heat losses. The matter will be treated as briefly as possible. From the chronological standpoint this type of gas turbine comes first. Actual turbines of this kind have been made, and tests of fuel consumption carried out. These have always been, as would be expected, heavy. In the case of the Karavodine turbine (*vide infra*) the fuel consumed per horse-power per hour amounted to 5 lbs. of gasoline. Taking the calorific value (not given) of the fuel to be 10,000 T.U., this gives an overall efficiency of only 3 per cent.

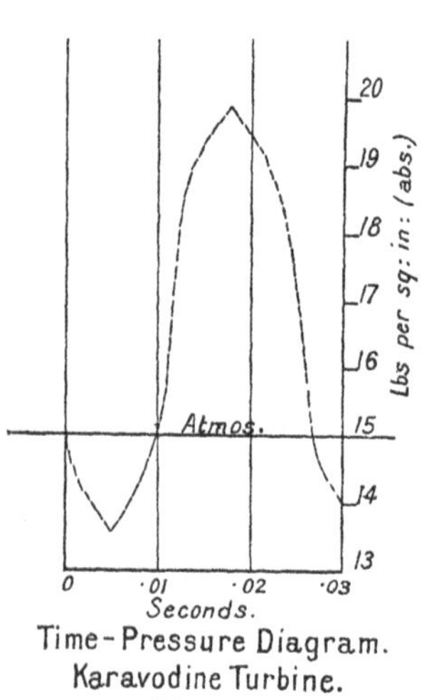

Time-Pressure Diagram.
Karavodine Turbine.
FIG. 55.

The Karavodine turbine used a wide combustion chamber with a restricted exit. Even then but a small rise in pressure was experienced. Fig. 55 shows a time-pressure graph taken from the Karavodine turbine. It will be noticed that the maximum pressure did not exceed 5 lbs. above the atmosphere.

The energy of the effluent gases from the chamber produced a suction effect; it will be seen from the diagram that this amounts to as much as 1 lb. below the atmosphere. This sufficed to raise the suction valve and admit the next charge. Once started, the action of the turbine was entirely automatic, and the presence of mechanically moved valves was entirely suspended. Were it possible to obtain anything approaching a reasonable efficiency, a turbine of this type has much to recommend it. In the first place, it presents, perhaps, the simplest, most compact, and cheapest form of prime mover in existence. For small powers fuel economy is a secondary object to compactness, simplicity and a small first cost. A small non-condensing steam engine only gives about 7 per cent. overall efficiency, and if it were possible to raise the efficiency of the Karavodine type of turbine to some such value, there would be many uses for it. A tentative design of such a machine is shown in Fig. 56; an inertia-pipe is shown coiled for compactness.

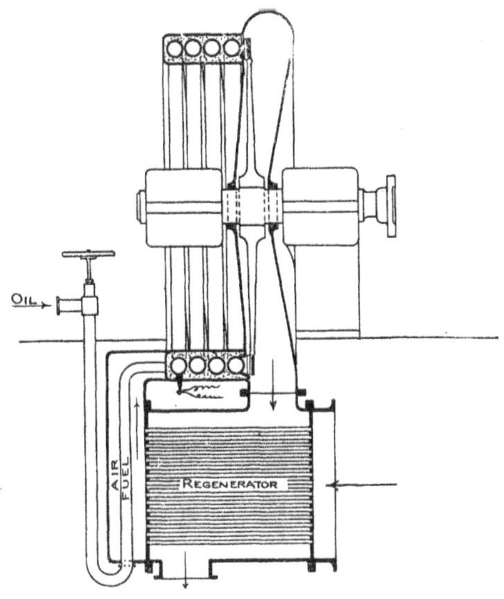

FIG. 56.—Explosion Turbine, Karavodine Type, with coiled "inertia" pipe and regenerator.

One or two points of theory in connection with the Karavodine turbine may be noticed in passing.

The system of setting successive columns of air in motion by the introduction and explosion of gas cartridges is, thermodynamically, very inefficient. Let m and V be the mass and velocity of the cartridge respectively; let M be the mass of the dilutant, great in comparison to m; then v, the final velocity of the total mass, is—

$$\frac{m}{\mathrm{M}+m}\mathrm{V}.$$

The kinetic energy of the cartridge $\mathrm{E} = \frac{1}{2} m\mathrm{V}^2$ and the subsequent kinetic energy of the total moving mass $\mathrm{E}_2 = \frac{1}{2} (\mathrm{M} + m) v^2$.

This equals—

$$\mathrm{E}_2 = \frac{1}{2}(\mathrm{M}+m)\frac{m^2}{(\mathrm{M}+m)^2}\mathrm{V}^2.$$

And—

$$\frac{E_2}{E_1} = \frac{m}{M + m}.$$

If M is great compared to m, the proportion of energy recoverable upon dilution is extremely small.

Air, however, is an elastic body; if it were perfectly elastic, and there were no loss of heat by conduction or radiation, there would not be the loss of energy noted above. The impulse on the element of air immediately in front of the charge would cause a momentary compression and corresponding expansion

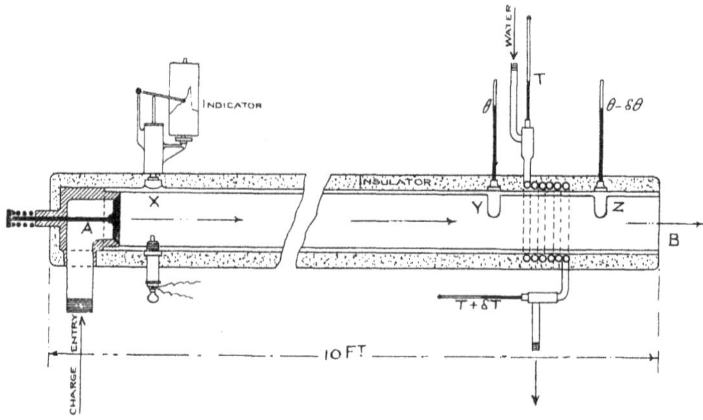

FIG. 57.—Experimental Apparatus for Determining Energy of Explosions in Open Vessels.

which would be translated throughout the whole mass of air in the tube, giving it a velocity, V′, such that—

$$mV^2 = (M + m)\,V'^2$$

and there should be theoretically no loss of energy in the system. Such, of course, is purely an ideal condition.

The question of the open-chamber explosion turbine depends entirely upon data furnished by experiment; it is unfortunate that more has not been done in this direction.

What is much wanted is an accurate determination of the *available* energy as compared to the heat put in, with explosions in open tubes. A simple and effective manner of carrying out such an investigation is here submitted by the writer to any interested enough to undertake the experiment; he has not himself had the opportunity of putting the method into practice.

Fig. 57 represents a long tube—10 feet or more—carefully lagged with some non-conducting material, to ensure, as far as possible, loss of heat by radiation. The tube is closed at one end, in which is placed a light suction valve. A recording pressure gauge (indicator) is placed at X. At Y and Z, at the other end of the tube, thermometers—or, better, recording pyrometers—are placed. Between the spots Y and Z is a small coil of pipe around the combustion tube through which water is circulated. The temperature of the water before passing through the coil, and also after having passed through the coil, is noted.

Let these temperatures be T and $T + \delta t$ respectively. Let the quantity of water passing per unit of time be q lbs. Let the temperature of the fluid at Y and Z be, respectively, θ and $\theta - \delta\theta$. During this time the explosions are at work continuously, fresh charges being automatically sucked in through the valve A and ignited by the sparking plug. The quantity of gas admitted per unit of time is measured (by meter or otherwise). Let this quantity be Q lbs.; let the cross-section of the pipe be A square feet. We have then—

$$C_p \cdot W \cdot \delta\theta = q \cdot \delta t,$$

where W equals the weight of working fluid passing per unit of time. If the mean density of the fluid equals ρ, and the area of the pipe section is A, we have—

$$C_p \cdot \rho V \cdot \delta\theta = q \cdot \delta t$$

where V = volume passing per second.

Therefore

$$C_p \cdot \rho A \cdot v \cdot \delta\theta = q \cdot \delta t,$$

where v is the velocity of the effluent gases.

Therefore

$$v = \frac{q \cdot \delta t}{C_p \cdot \rho \cdot A \cdot \delta\theta}.$$

Having obtained the actual velocity of the gas as it enters the turbine wheel, the work available is easily found, from the equation—

$$\text{Energy} = \tfrac{1}{2}\, mv^2/2g \,. = S.$$

We have the measurement of the heat supplied per unit of time—

$$H = Q \cdot h',$$

where h' is the heat value of the fuel.

Therefore, the overall efficiency of the turbine is given by—

$$\frac{S \cdot E_x}{H}.$$

where E_x is the thermodynamic efficiency of the turbine wheel. Thus all requisite knowledge for the efficiency of the turbine is obtained by the simplest of apparatus, except the value of E_x, which can readily be estimated, or some standard value may be taken.

It may not be altogether out of place to notice here one of the latest developments in speculation concerning gas turbine problems. This latest development—like almost all new ideas—is simply a reaction back to suggestions and schemes put forward years ago. The notion in question, emanating from the work of Mr. Humphrey, is simply that of applying the pump with which his name is associated to the raising of water for the driving of water turbines from the shafts of which the motive power is taken. Suggestions of this nature may be regarded as divisible into two classes: one in which the Humphrey pump is used to raise water to a reservoir from which the water turbines may be worked; the other in which a modified form of gas-explosion pump is used to set a column of water in a state either of continued motion in one direction, or of continued oscillation, the energy produced being, in either case, absorbed by a turbine wheel. A suggestion of this latter kind was put forward, at least by implication, in Newton's patent of 1856 (*vide* Chapter X.).

The former suggestion, namely, to raise water by the Humphrey pump and then utilise its potential energy by change into kinetic energy in a water turbine, may be examined as follows.

This involves, first of all, a consideration of the Humphrey pump in comparison to other methods of raising water on the ground of fuel economy.

In Dr. Unwin's official test carried out on the Humphrey pump and summarised in the minutes of the *Proceedings of the*

Institution of Mechanical Engineers for November, 1909, it is stated that the Humphrey pump gave a consumption of 83·12 cubic feet of gas per pump horse-power per hour, the calorific value of the gas being 147·29 B.T.U. per cubic foot. This gives a consumption of 12,230 B.T.U. per pump horse-power, or an overall efficiency, *exclusive of producer*, of 20·8 per cent. This is the best result reported. It is further stated in this same report, that "the fuel consumption in these trials, reckoned on work done in lifting water, was *less* than in any pumping arrangement, either by gas or steam, hitherto recorded."

It is unfortunate that a categoric statement of this nature should have been introduced into the report of the official tests, or into the summary thereof, printed in the *Proceedings* of the Institution.

Let us consider whether this categoric statement is correct.

In the test of three engines of the National Gas Engine Company by the Standards of Efficiency Committee of the Institution of Civil Engineers, in 1905, a "brake thermal efficiency" of 29·8 per cent. was obtained with one of the engines. The overall efficiency in the Humphrey pump was 20·8. We are, therefore, faced with the following inference: If the centrifugal pump (or piston pump and necessary gearing) gives a mechanical efficiency of anything equal to or over 20·8/29·8 = 69·8 per cent., then the statement in the report is categorically incorrect; assuming, that is, that the tests of the Standards of Efficiency Committee were reliable; and assuming that the producer efficiency is the same in either case.

If centrifugal pumps (or geared piston pumps) have *never* given an efficiency as high as 69·8 per cent., then the statement is sound.

Referring to Dr. Hopkinson and Mr. Chorlton's paper on the "Evolution and Present Development of the Turbine-Pump," as reported in the minutes of the *Proceedings of the Institution of Mechanical Engineers* for January, 1912, and the discussion thereon, evidence as to the efficiencies of centrifugal pumps was given even up to so high a figure as 85 per cent. In particular may be mentioned Mr. Patchell's evidence of a centrifugal pump giving an efficiency of 76 per cent. with a repetition of the same after a lapse of five years. It therefore follows that if the statement *re* the Humphrey pump above alluded to is correct, then the evidence, either of the Standards of Efficiency Committee, or that of Dr. Hopkinson, Mr. Patchell, Mr. Schubeler and others, is incorrect—a dilemma that leaves the earnest student of comparative efficiency in a sufficiently bewildered condition.

A comparison with the steam pumping engine may be made as follows. In the *Proceedings of the American Society of Mechanical Engineers* for December, 1906, the tests are given of a quadruple high-duty air compressor, by O. P. Hood. This engine gave a consumption of 189·7 B.T.U. per brake horse-power. This represents an overall efficiency of 25 per cent. Let the boiler efficiency be taken at 70 per cent., and the producer efficiency for the Humphrey pump at 80 per cent. (a basis of comparison tending to favour the latter), we have an overall efficiency for the complete Humphrey plant of 16·6 per cent., and an efficiency for the steam pumping plant of 17·5 per cent. If, therefore, the reciprocating pump gives a mechanical efficiency equal to or over 16·6/17·5 = 95 per cent., then there *are* examples of raising water by steam which are even more efficient than that of the Humphrey pump; provided, of course, the test here quoted is a reliable one, or that there have been other tests made, giving as good results. The student is more than ever at a loss to understand the statement that the tests on the Humphrey pump showed a "fuel consumption, reckoned in water lifted . . . *less than in any pumping arrangement, either by gas or steam, hitherto recorded.*"

It would seem, therefore, that a gas-engine driving a centrifugal, and the centrifugal driving a water turbine, would give as good (if not a better) overall efficiency as the Humphrey pump-driven turbine, and that, unless the Humphrey pump has large practical advantages to atone for its lack in efficiency (a doubtful matter—it may be remarked that each unit at the Chingford installation carries 200 valves, all operated by springs), the scheme is altogether out of court.

The final efficiency of the Humphrey pump water turbine plant may, however, be worked out, as a matter of abstract interest. The best impulse water turbines give from 78 to 84 per cent. efficiency (*vide* test results by Messrs. Escher, Wyss & Co.). Taking the gas producer efficiency at 84 per cent., and the higher value above quoted for the water turbine, and Dr. Unwin's test efficiency of 20·8 for the Humphrey pump, the final overall efficiency—

$$E_T = (20{\cdot}8 \times 84 \times 84) \text{ per cent.} = 14{\cdot}7 \text{ per cent.}$$

Taking the calorific value of coal at 7,000 T.U. per lb., this gives a coal consumption of 1·375 lbs. of coal per brake horse-power per hour. If this result is compared with the figures of coal consumption for other prime movers (*vide* p. 203), it will be seen that, for the steam-engine the values lie between 1·6 and 1·1, and those for the gas-engine between 1·16 and 0·86. Or it may be compared to the test of a National Gas Engine at Manchester, October 23rd, 1900, when a coal consumption of 0·822 lb. was obtained (*vide* Clerk, "Gas Engine," Vol. II., p. 83). This shows sufficiently plainly the impracticability of the scheme from the point of view of economy.

It may be argued, however, that, although on grounds of efficiency the application of the Humphrey pump for power transmission is not to be recommended, still, there are practical reasons for its adoption. It would seem, however, that there is little to recommend it on these grounds. In the Chingford installation, which was composed of five units, totalling in round numbers 1,000 H.P., the number of spring-operated valves amounted to something like 1,000 (200 per unit).

The Chingford plant occupies a space of 90 × 120, or 10,800 square feet for 1,000 B.H.P.; this is, of course, exclusive of the space that would be occupied by the water-turbine units. Herr Holzwarth's experimental gas-turbine plant occupied a space of 18·9 × 20 or 378 square feet, or about one-thirtieth the space for the same nominal horse-power. The mixed-fluid turbine plant shown in Fig. 31 occupies a space of 86 × 40, or 3,440 square feet, for 6,000 B.H.P., measured outside buildings. The excessive cumbersomeness of the Humphrey plant is evident.

A more reasonable proposition is that of setting a column of water in oscillation in what approximates to a U tube, the gas explosion chambers being at the ends of each arm of the U tube, and the water turbine contained in the bend of the U. A scheme of this nature has more to be said for it on the ground of space occupied than the direct application of the Humphrey pump in its present form. It is doubtful, however, whether its efficiency would be in any way superior to that of the Humphrey pump. The losses due to heat conduction from the explosion chamber, both by the water itself and by the metal sides of the vessel, would seem to be likely to be the same in either case, and these losses must be the principal heat losses experienced in the Humphrey pump. In the oscillatory type there would be an additional loss in the water turbine owing to the variation of the velocity of the water through the turbine, a loss similar to that experienced with the explosion-gas turbine, which, in the case considered, in connection with the Holzwarth turbine, amounted to some 18 per cent. (*vide* p. 125).

The suggestion that a continuous column or "endless belt" of water should be maintained in a continuous state of flow by a series of gaseous explosions, and that the kinetic energy of the water thus produced should be absorbed by impulse wheels, seems to offer the most likely field for successful investigation. The problem, however, presents several difficulties; it is difficult to see how the variation in velocity could be overcome without the intervention of a reservoir and the accompanying conversion of kinetic into potential energy, in which case conditions analogous to those of the direct application of the Humphrey pump would be inevitably set up.

In so speculative a field as this it is dangerous to do other than generalise, but one advantage that any scheme by which the kinetic energy of water is used as an intermediary agent between the energy of gaseous expansion and its conversion into rotational energy upon a turbine shaft may be noted. This is the really immense advantage of unlimited speed control. The gas turbine of this type would, at least, be no longer confined to the driving of high-speed dynamos, but would be directly applicable to the driving of slow-speed machinery. At least, it may be admitted that this is no small advantage.

CHAPTER VI. THE VARIATION IN THERMODYNAMIC CONSTANTS AND ITS EFFECT UPON EFFICIENCY

IN the foregoing chapters the various efficiencies have, in all cases, been worked out on the assumption of certain arbitrary values for the specific heats and other constants employed in the calculations. This procedure was deemed more sane and more logical than to overburden every calculation with the elaborate paraphernalia of specific heats and specific volumes varying with the composition of the mixture (which is dependent upon the temperature in the combustion chamber) and of specific heats varying with actual temperature rise, according to the evidence afforded by Messrs. Holborn and Henning, Dugald Clerk, Langen, Mallard, and Le Chatelier and others.

With German writers (and the practice is by no means confined to Germany) it has been the custom to begin with the most elaborate consideration of all possible changes in thermodynamic constants, to drag these into every calculation of efficiency, on the ground—apparently—that the only logical method is to adopt the absolute values for thermodynamic constants at the very start, no matter how small is the difference in efficiency which is produced from that obtained from the use of the arbitrary values of common practice. Such a procedure may be applicable to a history of ethics, but it is inefficient in engineering; it is reminiscent of the mediæval historian who, when about to write a history of England, found it almost invariably necessary to begin with the siege of Troy—on the grounds of logical sequence.

It must be remembered that the difference in efficiency obtained by taking into account the changes in specific heat is very small. In the estimation of the overall efficiency of the gas turbine it is necessary to use arbitrary constants for the thermodynamic efficiencies of the turbine and pump, which may vary some 10 per cent. in any one machine. It is obviously inefficient to trouble about an error of 1 per cent. (or less) in cases where a probable error of 10 per cent. is introduced voluntarily. It is even a greater waste of time so to do when the factors that determine the value of the small error above alluded to are so indeterminate as to admit of 50 per cent. variations between contemporary experimenters.

It remains to be seen how small the error produced by neglecting the changes produced in the thermodynamic constants is.

Consider a sample of producer gas of a particular composition. The one here chosen is taken from Robinson's "Gas Engine," p. 586. The sample is a sample of Mond gas (without ammonia recovery) and its composition is as follows :—

COMPOSITION BY VOLUME.

	Cubic Feet.
Hydrogen, H_2	20·5
Methane, CH_4	2·3
Carbon monoxide, CO	15·7
Olefines	traces.
Carbon dioxide, CO_2	11·7
Nitrogen, N_2	49·8
	100
Air added	108
Total	208

We may neglect the small percentage of methane by increasing the proportion of hydrogen by the same amount. This simplifies calculations later.

We now have—

COMPOSITION BY VOLUME.

	Cubic Feet.
H_2	22·8
CO	15·7
CO_2	11·7
N_2	49·8
Air	108·0
	208·0

or,

COMPOSITION BY WEIGHT.

	lbs./cub. ft.	lbs.
H_2 . . .	·00561	·1278
CO . . .	·07800	1·2240
CO_2 . . .	·12340	1·4430
N_2 . . .	·07810	3·8900
Air . . .	·07670	8·2800
		14·9648

This is the composition before ignition. We have now to obtain the composition by weight of the mixture *after* ignition. We have the following reactions taking place—

$$2H_2 + O_2 = 2H_2O$$
$$\cdot 1278 + 1\cdot 02 = 1\cdot 1478$$

and

$$2CO + O_2 = 2CO_2$$
$$\cdot 1224 + \cdot 7 = 1\cdot 924$$

and the composition by weight of the working fluid becomes—

	lbs.
H_2O	1·1478
CO_2	3·3670
N_2	9·6500
Air (remainder) . . .	·8000
	14·9648

Consider 1 lb. of gas *before* combustion :—

COMPOSITION BY WEIGHT.

	lbs.
Hydrogen	·0192
Carbon monoxide . . .	·1840
Carbon dioxide . . .	·2158
Nitrogen	·5810
	1·000

Air added, 1·24 lbs.

Consider 1 lb. of mixture *after* combustion :—

COMPOSITION BY WEIGHT.

	lbs.
Steam	·1720
Carbon dioxide . . .	·5040
Nitrogen	1·4440
Air (remainder) . . .	·1200
Total .	2·2400

The specific heat of the mixture has now to be obtained from those of its constituents—

(A) SPECIFIC HEAT OF MIXTURE BEFORE COMBUSTION.

Constituent.	Composition by weight. (W).	C_p.	C_v.	C_p . W.	C_v . W.
Hydrogen . .	·0192	3·400	2·400	·0653	·0461
Carbon monoxide .	·1840	0·242	0·173	·0445	·0318
Carbon dioxide .	·2158	0·210	0·165	·0453	·0356
Nitrogen . .	·5810	0·235	0·175	·1364	·1017
	1·0000	—	—	·2915	·2152
Air . . .	1·2400	—	—	·2980	·2115
Mixture . .	2·2400	0·263	0·190	·5895	·4267

Therefore C_p for mixture *before* composition $= 0·263$
Therefore C_v ,, ,, ,, $= 0·190$
Therefore γ ,, ,, ,, $= 1·384$

(B) SPECIFIC HEAT OF MIXTURE AFTER COMBUSTION.

Constituent.	Composition by weight. (W).	C_p.	C_v.	C_p . W.	C_v . W.
Steam . . .	0·1720	·480	·355	·0826	·0611
Carbon dioxide .	0·5040	·210	·168	·1058	·0832
Nitrogen . .	1·4440	·235	·175	·3394	·2530
Air (remainder) .	0·1200	·240	·172	·0288	·0206
Mixture . .	2·24	·248	·787	·5566	·4179

Therefore C_p for mixture *after* combustion = 0·248
Therefore, C_v ,, ,, ,, = 0·187
Therefore, γ ,, ,, ,, = 1·330

The calorific value of the gas in question is obtained from an examination of the heat evolved by its constituents.

1 lb. of hydrogen gives on combination 34,170 T.U.
Then ·0192 lb. ,, ,, 655 ,,
1 lb. of carbon monoxide ,, 2,400 ,,
Then ·184 lb. ,, ,, 442 ,,
Therefore, 1 lb. of gas has thermal value of, H = 1,097 T.U.

Thus we have the values of C_p and C_v and γ for the combustion mixture composed of the gas in question together with the minimum proportion of air necessary for complete combustion. For a mixture of 1 lb. of gas with 1·24 lbs. of air—

	Before Combustion.	After Combustion.
C_p . .	·263	·248
C_v . .	·190	·187
γ . .	1·384	1·330

Assuming the values for C_p and C_v to remain constant with the increase in temperature, the temperatures produced upon combustion with this mixture are—

T = 1,980° C. heat absorbed at constant pressure.
T = 2,635° C. heat absorbed at constant volume.

As these temperatures are not allowable in practice, the mixture must be further diluted with air. This dilution, of course, changes the values of C_p, C_v and γ for the mixture, as the composition of the mixture is changed. The effects of this dilution on the specific heats and the maximum temperature are shown in Table II. So far the change in the specific heats, due to the rise in temperature, have been neglected.

The effect of this change in the value of C_p and C_v is shown (not to scale) in the entropy-temperature diagram (Fig. 58), in which nm_1, nm_2, nm_3, etc., represent the heat absorption curves for mixtures, 1, 2, 3, etc. The effect on the heat absorption curve by the change in the value of C_p, C_v by change of mixture is, of course, here, greatly exaggerated; its actual value on the entropy diagram is shown in Fig. 62 (*vide infra*), where the heat absorption curves for the varying mixtures all lie within the shaded area between the inner and outer heat absorption lines.

There is a change in the specific heats produced when ignition takes place; this, however, does not affect the efficiency of the system. In Fig. 59 is shown the heat absorption lines on the entropy-temperature diagram (not to scale) for the

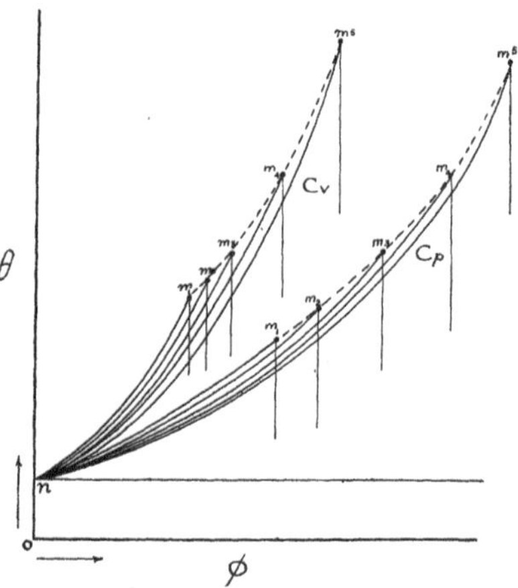

FIG. 58.—"ΘΦ" Diagram, showing Changes in Heat Absorption Lines for varying Composition of Mixture. (Not to scale.)

values of C_p and C_v before and after combustion; *cb*, *b'* representing heat absorption lines for values of C_p, C_v before combustion; *ca*, *a'*, for values of same after combustion. If combination does not take place until the temperature θ_1 is reached, the area *abk* will project beyond the adiabatic *ae*, and a fresh adiabatic will have to be dropped from *b*, *b'*, thereby effecting the efficiency of the cycle. If combination begins to take place at a lower temperature, a smaller portion of the area between the two curves will project beyond the adiabatic from *a*, and the effect on the thermal efficiency will be still smaller. In point of fact, however, the combination of the gases begins

Table II.

Mixture.		Air added to combustion Mixture. (1). ω.	Weight of Mixture. W.	Before ignition.		After ignition.		Before.	After.	T'_{cp} (°C) T'_{cv}		Composition.
Gas.	Air.			$C_p = \frac{\cdot 5895 + \cdot 24\omega}{W}$	$C_v = \frac{\cdot 4267 + \cdot 172\omega}{W}$	$C_p = \frac{\cdot 5566 + \cdot 24\omega}{W}$	$C_v = \frac{\cdot 4179 + \cdot 172\omega}{W}$	$\gamma = \frac{C_p}{C_v}$	$\gamma = \frac{C_p}{C_v}$	$= \frac{H}{C_p . W} + t$	$= \frac{H}{C_v . W} + t$	Weight of gas per unit weight of mixture.
										(H = 1097) $t = 15°$.		per cent.
1	1·24	0	2·24	·263	·190	·248	·187	1·384	1·330	1,980°	2,635°	44·7
1	1·74	·5	2·74	·259	·187	·247	·184	1·382	1·341	1,655	2,190	36·5
1	2·24	·1	3·24	·256	·185	·246	·182	1·383	1·351	1,392	1,875	30·8
1	2·74	1·5	3·74	·254	·183	·245	·181	1·387	1·353	1,213	1,635	26·7
1	3·24	2	4·24	·252	·182	·244	·180	1·387	1·354	1,075	1,455	23·6
1	3·74	2·5	4·74	·251	·181	·244	·179	1·388	1·367	965	1,308	21·1
1	4·24	3	5·24	·250	·180	·244	·178	1·390	1·370	873	1,193	19·1
1	4·74	3·5	5·74	·249	·179	·243	·178	1·390	1·364	802	1,087	17·4
1	5·24	4	6·24	·2486	·1786	·243	·1772	1·392	1·370	740	1,008	16·0

to take place very early in the process of heat absorption and the curve *cdb* is soon deflected towards the curve *cfa*, in some such form as the curve *cda*, lying entirely within the adiabatic dropped from *a*. This means that the change in specific heats before and after combustion does not affect the efficiency of the cycle provided the values of C_p and C_v be taken as those for the mixture after combustion.

The results of the values of the specific heats, given in Table II., p. 143, are shown graphically in the diagram, Fig. 60. The values of C_p, C_v and γ are plotted against the temperature produced (° C.). This is not concerned with the increase in specific heat due to *temperature.* It shows the change in these constants due to the change in the composition of the mixture required to produce certain maximum temperatures. For example—to produce a temperature of 1,700° C. (under conditions of constant volume) a mixture containing 28 per cent. by weight of Mond gas is required. This relation is shown by the line G. The value of C_v for such a mixture is ·1813 after combustion (line A), and ·1837 before (line B). The value of γ is 1·352 before (line C), and 1·387 after (line D). C_p is ·2454 before (line E), and ·255 after (line F). It will be noticed that there are corresponding lines, giving corresponding values for the same constants, in the case of heat absorption at constant pressure (viz., lines *a, b, c, d, e, f, g*). If, for example, the temperature of 1,700° C. has to be reached by heat absorption at constant pressure, a richer mixture is necessitated ; a mixture, namely, of 38 per cent. of gas. This alters

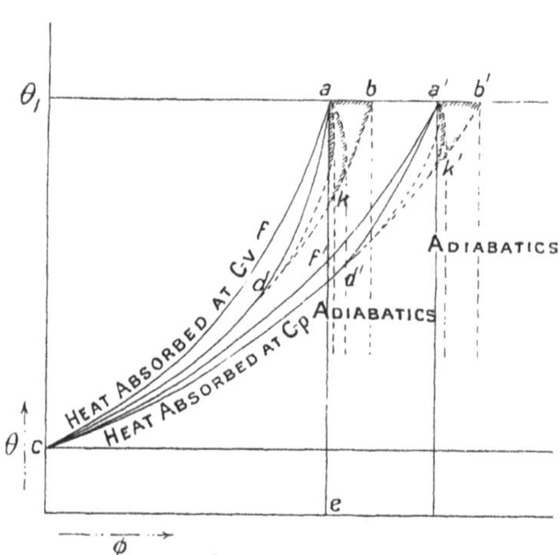

Fig. 59.—"ΘΦ" Diagram, showing Change in Heat Absorption Lines for Mixture before and after Combustion. (Not to scale.)

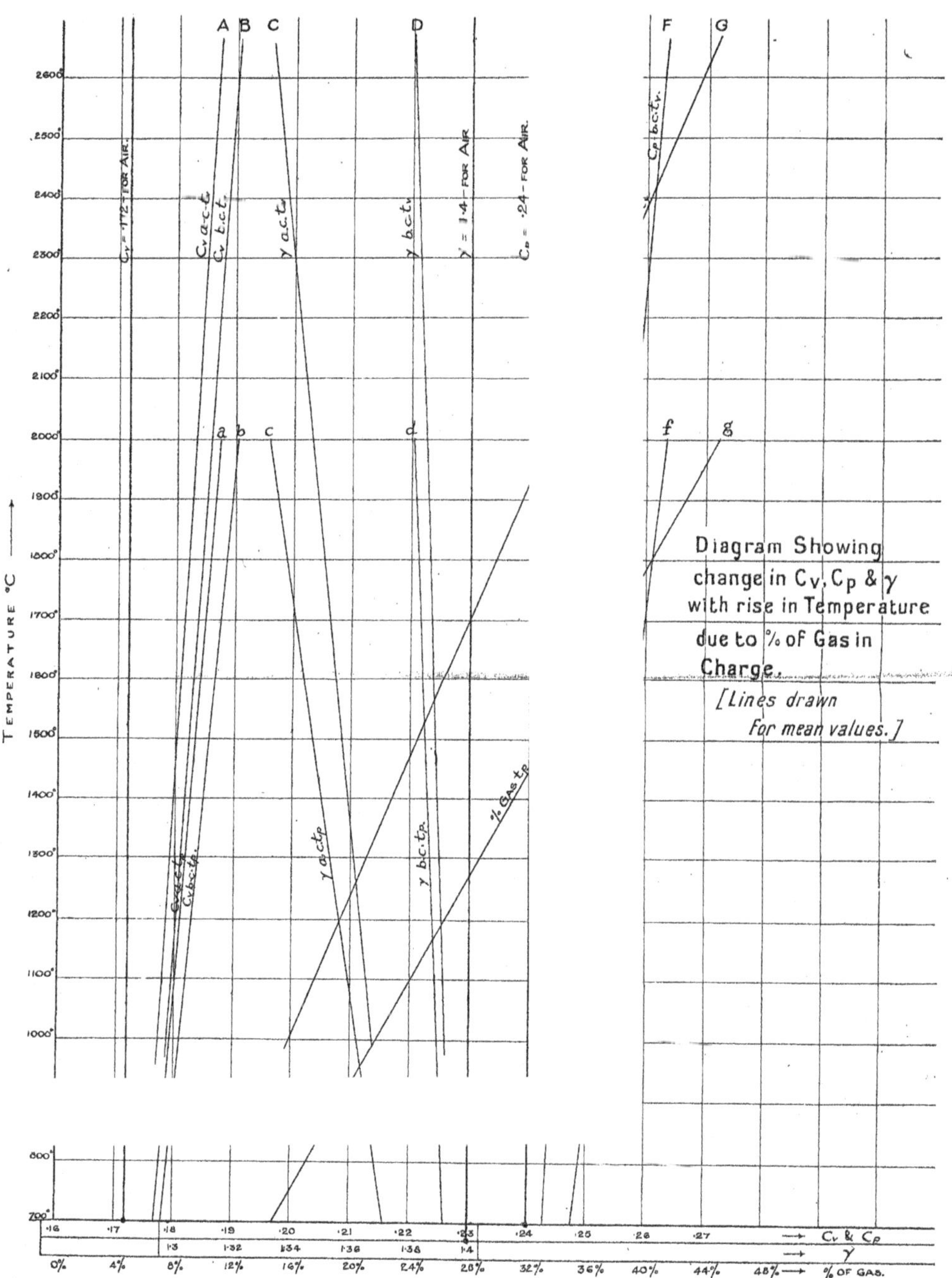

FIG. 60.—Lines A, B, C, D, E, F, G, for Heat Absorption at Constant Volumes. Lines *a*, *b*, *c*, *d*, *e*, *f*, *g*, for Heat Absorption at Constant Pressure.

throughout the values of the other constants. Thus, the value of C_v after combustion becomes ·1848 (line *a*) instead of ·1813. Similar changes occur in the other constants, their values being found from the lines *a, b, c, d, e, f*.

THE INCREASE OF SPECIFIC HEATS WITH TEMPERATURE.

In the report of the Committee on Gaseous Explosions of the British Association (Dublin, Section G., 1908 *et seq.*) the

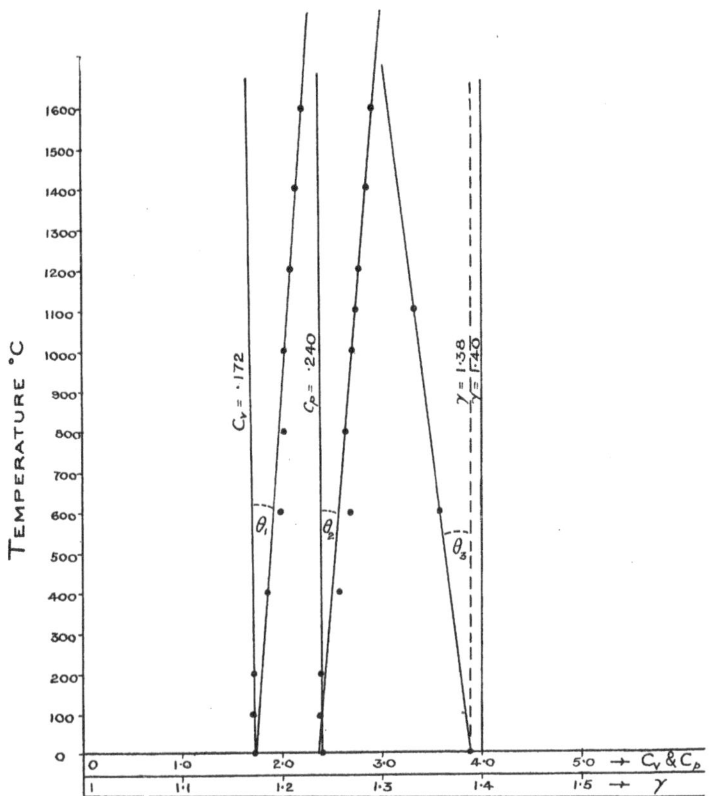

Fig. 61.—Diagram showing Variation in Values of C_v, C_p, and γ, with Rise of Temperature. (From data of B.A. Report, 1908.)

question of the actual change in the value of C_p and C_v and γ with change in temperature is very fully discussed. The data of various experimenters differ considerably, and the difficulties consequent upon the determination of specific heats of gases at high temperatures are very great. The writer has taken the curve shown on p. 22 of the 1908 report, which gives a mean between the values of Messrs. Holborn and Henning, Dugald Clerk, Langen and Mallard and Le Chatelier, interpreting it, however, in the form of a straight line, giving a line equation for the change in these constants with temperature. In the present unsettled state of the question such is not too drastic a liberty, and the error, even if the curve is the correct reading, is extremely small. The value of this increase is shown in Fig. 61, for C_p, C_v and γ, giving—

$$C_p = \alpha + \beta\theta \quad \text{where} \quad \alpha = \cdot 24$$
$$\beta = \cdot 0000337$$
$$C_v = \mu + \nu\theta \qquad \mu = \cdot 172$$
$$\nu = \cdot 0000356$$
$$\gamma = \lambda + \kappa\theta \qquad \lambda = 1 \cdot 38$$
$$\kappa = \cdot 000052$$

The effect of this change in the specific heats on the " $\theta\phi$ " diagram is to shift the heat absorption lines towards the right. The heat absorption line being now obtained by the formula—

$$\phi_1 = (\alpha + \beta\theta) \int \frac{d\theta}{\theta}$$
$$\phi_2 = (\mu + \nu\theta) \int \frac{d\theta}{\theta}.$$

The heat rejection lines, are, of course, determined in the same way.

The temperature-entropy diagram for the variable specific heats is shown in Fig. 62. The curves KA and KA′ represent the heat absorption at constant volume and constant pressure, respectively, for a perfect gas. The curves KB, KC enclose the area (shaded) which contains the heat absorption curves for varying mixtures according to maximum temperature required (*vide* Fig. 58). The curves KD and KD′ represent the heat absorption at constant volume and constant pressure, with the specific heats varying. The cycles are completed by dropping adiabatics upon the horizontal representing the maximum value of the temperature after expansion, and completing the diagram in the usual way.

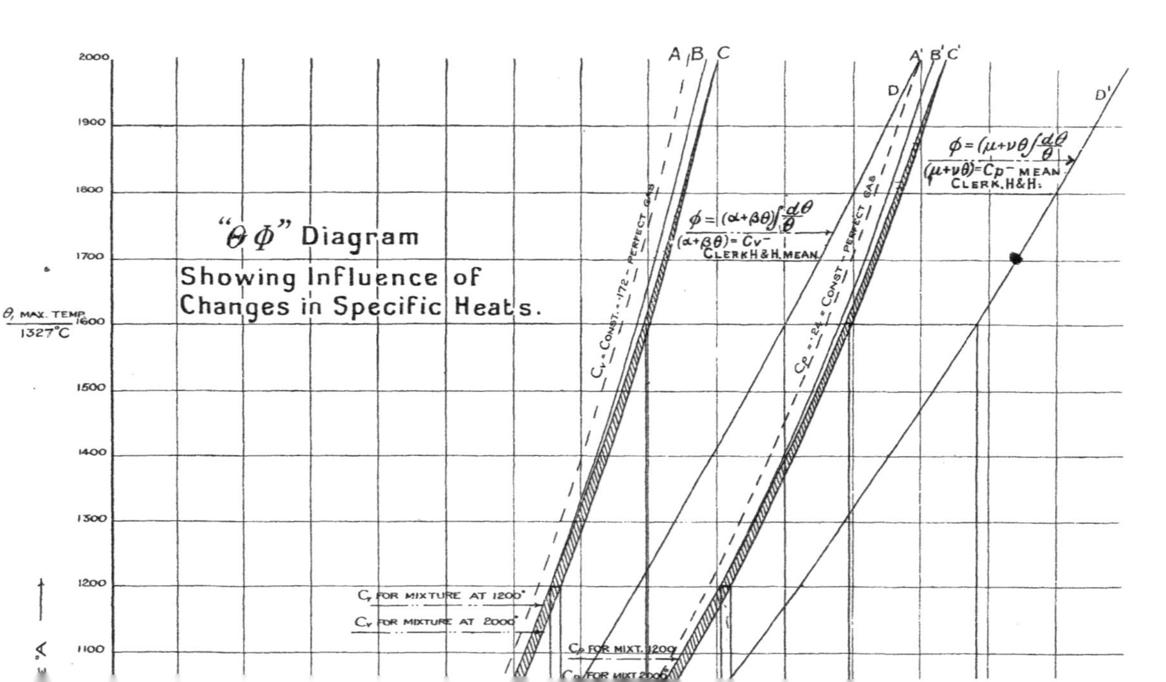

"θ Φ" Diagram
Showing Influence of
Changes in Specific Heats.
A B C
A' B' C'
D
D'
Cv = Const. = ·172 – Perfect Gas
Cp = ·24 = Const – Perfect Gas
φ = (α+βθ)∫dθ/θ
(α+βθ) = Cv – Clerk H & H. Mean
φ = (μ+νθ)∫dθ/θ
(μ+νθ) = Cp – Mean Clerk. H&H.
Cv for mixture at 1200°
Cv for mixture at 2000°
Cp for mixt. 1200
Cp for mixt 2000
2000
1900
1800
1700
1600
1500
1400
1300
1200
1100
θ, Max. Temp. 1327°C

θ_2 TEMP. ON BLADES 773
500°C
θ' FROM REGR. k
215°C
800
700
600
500
400
300
K
200
100
0
WORK OF PUMP
HEAT REJECTED FROM REGENERATOR
0
·1
·2
·3
·4
·5
ENTROPY. ⟶

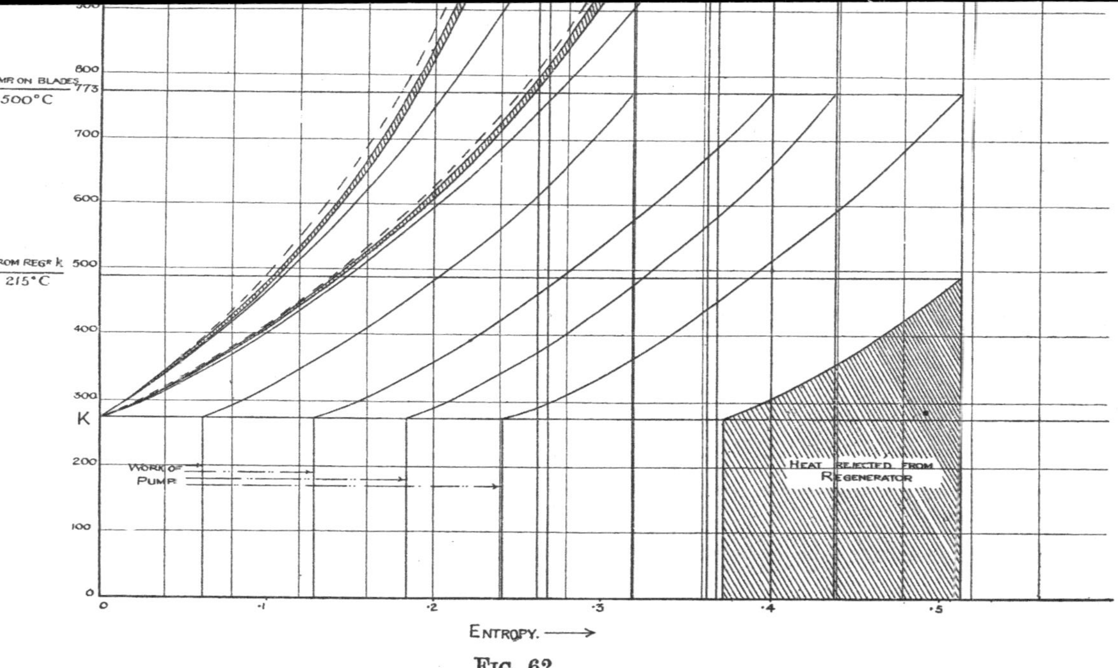

FIG. 62.

It will be noticed from the above diagram that, with the variable specific heats, the work done on the pump is increased, as compared with the similar case for the constant specific heats ; the temperatures θ_1 and θ_2 remaining constant.

COMPARISON OF EFFICIENCIES.

It is now to be seen what is the effect of the changes in the constants above discussed upon the calculation of the thermal efficiencies of the gas turbine as compared to those already taken in the preceding chapters.

Let an actual example be chosen.

Constant-pressure Single-fluid Gas Turbine.

Fixed ; temperature after explosion, $\theta_2 = 773°$ A. (500° C.) ; expansion ratio, R = 10. The initial temperature, θ_1, will be about 1,473° A. For the requisite mixture to produce such a temperature under conditions of heat absorption at constant pressure—

$$C_p \text{ at } 15° \text{ C.} = \cdot 245,$$
$$\gamma \text{ at } 15° \text{ C.} = 1{,}360$$

(*vide* Fig. 61).

The mean value of γ between θ_1 and θ_2, from diagram (Fig. 61) is 1·32. This gives a value for $\frac{\gamma - 1}{\gamma}$ of ·243 approximating slightly higher (actual value ·242) on account of the fact that the actual value of θ_1 is slightly less than 1,473° A.

We have now—

$$\theta_1 = \theta_2 \,.\, R^{\cdot 243}$$
$$= 1{,}360° \text{ A},$$

R being equal to 10.

Now work done by the working fluid in the turbine—

$$W = C_p\ (\theta_1 - \theta_2),$$

and C_p has here to be the mean value for that constant between 1,360° and 773° A. Therefore—

$$W = \cdot 271\ (1360 - 773)$$
$$= 160 \text{ T.U.}$$

Consider the question of regeneration. Let the exit temperature from the regenerator be, as before, fixed at 215° C. Then heat lost from waste gases—

$$h = C_p(200)$$

where C_p is the mean value between 215° and 15° C.

Therefore—

$$h = \cdot 248(200)$$
$$= 49{\cdot}6 \text{ T.U.}$$

Consider work done by pump. We have work done by pump—

$$w = p_0 v_0 \log_\epsilon R.$$

This equation involves the factor of volume. We have, in fact, to find the exact volume occupied at 15° C. by 1 lb. of the working fluid under examination after combustion. We have composition by weight—

Steam	·172
Carbon dioxide . .	·504
Nitrogen . . .	1·444
Air	·120
	2·240

The steam, however, is condensed in the condenser, and has not to be pumped out by the pump. We have, therefore, only 2·068 lbs. of fluid to deal with.

For an initial temperature of about 1,200° C., 1·5 lbs. of air are required as dilutant to the above mixture.

Thus we have—

		Cub. ft./lb.	Volume.
Steam	·172	—	—
Carbon monoxide . .	·504	8·1	4·1
Nitrogen . . .	1·444	12·8	18·5
Air	1·62	12·4	20·1
Mixture	3·740	—	42·7

Therefore, volume per lb. of mixture for pump—

$$v_0 = 11{\cdot}4.$$

Then, work done by the pump—

$$\begin{aligned} w &= 14{\cdot}7 \times 144 \times 11{\cdot}4 \times \log_e 10 \quad \text{ft./lbs.} \\ &= 17{\cdot}25 \log_e 10 \quad \text{T.U.} \end{aligned}$$

This gives a value for w of 39·6 T.U.
Now the thermal efficiency—

$$\begin{aligned} \eta &= \frac{W - w}{W + h} \\ &= \frac{160 - 39{\cdot}6}{160 + 49{\cdot}6} \\ &= 57{\cdot}6 \text{ per cent.} \end{aligned}$$

Compare this with the result obtained by the use of the arbitrary values of the thermodynamic constants, assuming C_p, C_v, and γ to be constant throughout the range of temperature considered. Let C_p be constant at ·25 : γ, at 1·38.

$$\begin{aligned} \theta_1 &= 773 \,.\, R^{\cdot 275} \\ &= 1{,}457^\circ \text{ A.} \end{aligned}$$

Therefore

$$\begin{aligned} W &= {\cdot}25\,(1{,}457 - 773) \\ &= 171 \text{ T.U.} \end{aligned}$$

Heat lost from regenerator—

$$h = {\cdot}25 \times 200 = 50 \text{ T.U.}$$

Negative work on pump—

$$\begin{aligned} w &= 18{\cdot}7 \log 10 \\ &= 43 \text{ T.U.} \end{aligned}$$

Therefore, we have—

$$\begin{aligned} \eta &= \frac{W - w}{W + h} \\ &= \frac{171 - 43}{171 + 50} \\ &= \frac{128}{221} = 57{\cdot}9 \text{ per cent.} \end{aligned}$$

It is thus seen that, neglecting the change in thermodynamic constants completely, and assuming the orthodox values for the same—as has been done in the preceding chapters—the value of η in this particular case comes to ·579. Taking into account all the possible variations in these constants, and working out the thermal efficiency from the most absolute basis, the value of η in this same case comes to ·576. That is to say, the values obtained for the thermal efficiencies have so far been 0·3 per cent. too high. On the overall efficiency this would tend to become less.

The author feels that it would have been useless and inefficient to have burdened the preceding chapters with the matter that has been touched upon in this chapter for the sake of an "error" of less than 1 in 300, more particularly when the grounds upon which that "error" is based are of so uncertain a nature. In succeeding chapters the arbitrary constant values of C_p, C_v and γ will again be adopted (namely, ·25, ·18 and 1·38).

BOOK II. THE PRACTICE

CHAPTER VII. ACCESSORY MACHINERY

It has been said with considerable truth that the success of the gas turbine as a prime mover rests altogether with the rotary pump. This is certainly so in the case of the constant-pressure, single-fluid turbine ; it is less so with the mixed-fluid, constant-pressure turbine ; it barely applies to the case of the constant-volume turbine at all. As such future that there may be for the gas turbine as a heat engine, at least in the opinion of the writer, rests rather with the constant-pressure than with the constant-volume types, the efficiency of the rotary pump may be considered to have a very weighty influence upon the elucidation of the gas turbine problem.

But there are other accessory matters—points in construction, methods in design—that are almost as important as the question of the rotary pump. For convenience of treatment these matters have been divided up in the following manner—

(A) The producer and furnace.
(B) The turbine wheel ; water-cooling.
(C) The regenerator and condenser ; the sulphuric acid question.
(D) Governing.
(E) The pump.

It is of these subjects that the present chapter treats.

(A) THE PRODUCER AND FURNACE.

It may be as well, to begin with, to dismiss a fallacy that has from time to time been suggested in connection with the gas turbine. This is the use of producers working " under pressure," thereby delivering the fuel to the combustion chamber at the maximum pressure of the cycle. The matter does not apply, of course, in cases where the maximum pressure is that of the atmosphere. Apart from any practical difficulty of feeding the coal to the producer under super-atmospheric conditions (the great difficulty that presented itself in the construction of the Buckett air engine), it must be remembered that the air fed to the producer must still be compressed to the required pressure, and also that the air added to the gas in the combustion chamber to support combustion and produce the necessary dilution requires the same ratio of compression. The ratio of air to producer gas necessary to produce the limit of maximum temperature required, is about 3 to 1 (*vide* Table II.). This means that even were it possible to avoid compressing the feed to the producer, only 25 per cent. of the work on the pump would be avoided. As the air supplied to the producer has to be compressed also, this apparent saving in work disappears. It is only in one case—that of the constant pressure, super-atmospheric *oil* turbine—that any saving can be effected by initial compression of the fuel. For the production of the necessary initial temperature, 1 lb. of oil requires 32 lbs. of air for combustion and the necessary dilution. Neglecting the work done in pumping the *liquid* fuel into the combustion chamber, there is a saving of 3 per cent. on the work done by the pump as compared to the same case with gas fuel. Only one advantage is it possible to obtain from using producers under pressure, and this is the ability to use the gas from the producer in the combustion chamber without initial cooling.

This abolition of the heat losses in the cooler is an important factor in gas turbine economy. It is easy, of course, in the case of the sub-atmospheric gas turbine to burn the gases straight from the gas producer without any previous cooling.

Such a plan is not permissible in the case of the gas engine, on account of the cooling effect of the walls of the cylinder and the subsequent deposition of tar and other foreign matter were such not removed in the cooler and scrubber. With the gas turbine, however, or, that is, with the constant-pressure gas turbine—for with the explosion gas turbine the same difficulties would occur as in the case of a gas engine cylinder—the gas is thoroughly burnt in an open and roomy furnace before entry into the turbine wheel. It may, therefore, be assumed that all the tarry matter would be thoroughly burnt in the furnace before entering the turbine.

In a case given of a Mond gas plant (Robinson) the gas left the producer at a temperature of 450° C., and the cooling tower at 50° C. Taking the value of C_p for the gas to be ·263 (*vide* p. 141), this gives a loss upon scrubbing and cooling of 8 per cent. of the heat value of the gas. Taking the efficiency of the producer to be 84 per cent., this raises the efficiency to a value of 92 per cent. Or, in a general comparison with the gas engine, an advantage in favour of the gas turbine of some 8 per cent. in fuel consumption.

It is, however, reasonably objected that to do away with the scrubber would go to the production of dust in the turbine, which would be highly prejudicial to the blades and bearings. To avoid this defect a system of "dry scrubbing" has been suggested, by which the dust and such foreign solid matter is removed from the gas without occasioning any appreciable lowering of its temperature. How far such a system would be successful it is difficult to say; the writer does not know of any instance in which "dry scrubbing" has been applied.

The design of furnace is a matter of extreme importance to the successful working of the gas turbine. It is convenient to consider the furnaces (combustion chambers) for three types of gas turbines: the constant-pressure, single-fluid gas turbine; the mixed-fluid, constant-pressure gas turbine; and the explosion gas turbine.

(I) The Furnace for the Single-fluid, Constant-pressure Gas Turbine.

The ideals that have to be aimed at in this case are the prevention of loss of heat by radiation and conduction and the durability of the combustion chamber itself. It should, of course, be theoretically possible to use a metal chamber, uninsulated by any non-conducting material, and so to construct the chamber that by the continual flow of the supply to the chamber around the chamber there should be no loss of heat save that due to the difference of temperature for heat transmission through the walls of the vessel. Such a process is a regenerative one; a section of a combustion chamber designed for this purpose is shown, diagrammatically, in Fig. 63. Owing, however, to the low coefficient of conductivity of gases, such a scheme is not possible in practice. There are three losses of heat possible from the combustion chamber: loss due to conduction; loss due to convention; loss due to radiation; we are really only concerned with the first and the last of these losses.

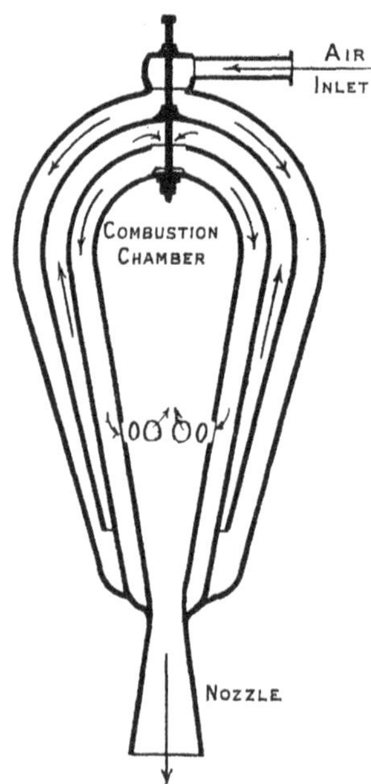

FIG. 63.—Regenerative Constant-Pressure Combustion Chamber.

The loss due to conduction is dependent on the material of which the chamber is composed. It is proportional to the temperature difference between the inside and the outside of the walls of the chamber. The loss due to radiation is proportional to the fourth power of the difference in temperature between the outside of the combustion chamber walls and the surrounding medium. It will be thus seen, that if the latter loss is at all great, an increase in the value of the initial temperature must greatly reduce the efficiency of the combustion chamber (*vide* Fig. 37).

The type of combustion chamber used in the Armengaud and Lemale gas turbine is shown in Fig. 64. P is the entrance for fuel (in this case oil) which is gasified in the "atomiser" D. Air is admitted at A. Cold water is circulated

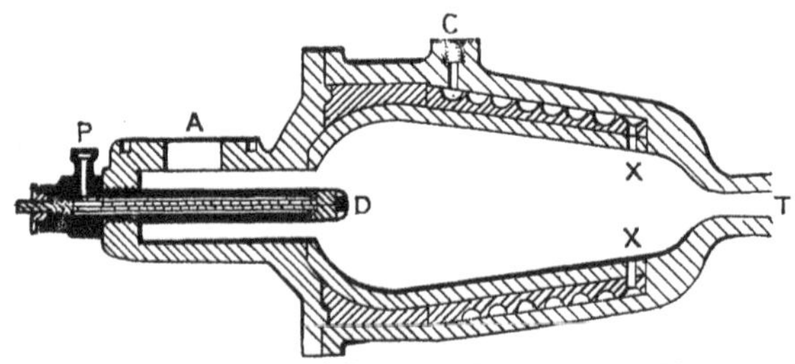

FIG. 64.—Armengaud-Lemale Combustion Chamber.

around the chamber at C., vaporised in the process, and the steam produced added to the working fluid through the ports

X. The fluid enters the expanding nozzle at T. It will be seen that such a combustion chamber exactly corresponds to that for a super-atmospheric steam-and-air turbine, the only difference between the two being the quantity of water fed into the combustion chamber. In the Armengaud-Lemale turbine this quantity seemed to be small, the water merely being used for water-jacketing purposes. A section of the "atomiser" is shown in Fig. 65. The fuel is fed through the annular space B, and enters the combustion chamber at C; it is thus deflected backwards against the entering stream of air, and a cone of inflammable vapour is formed around the end of the atomiser. The mixture is ignited by the platinum wire at P, which is protected by the steel cap S. An insulated rod passes down the centre of the atomiser to the necessary electrical connections. The combustion chamber of the Armengaud and Lemale turbine was lined with carborundum, a matter that led to disaster in working. The radiation losses from the chamber must have been considerable.

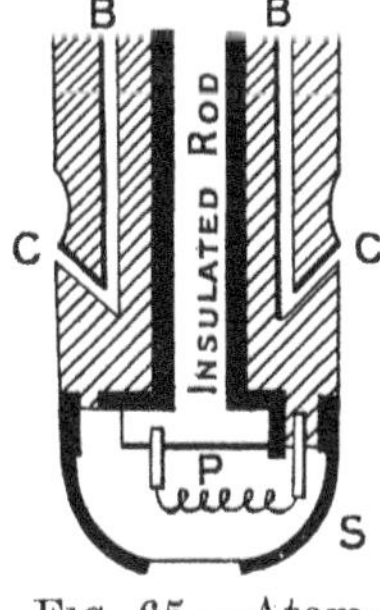

FIG. 65.—Atomiser.

A scheme for using the steam raised from the water jacket

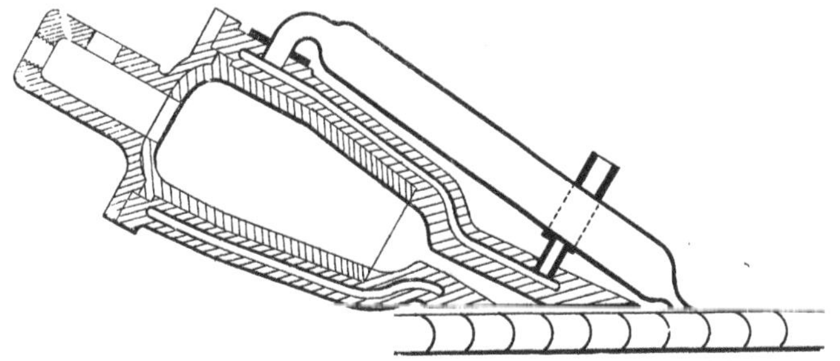

FIG. 66.—Armengaud-Lemale Combustion Chamber with separate Steam Expansion.

in a separate nozzle is shown in Fig. 66. The chief difficulty in the construction of the combustion chamber for this type of gas turbine is the inability to use efficient water cooling, on account of the excess of steam raised, and the subsequent loss from the walls of the chamber by radiation and conduction.

This may be, and, indeed, must be, remedied by some system of lining the combustion chamber with protecting material, as the temperatures likely to be involved would be too great for a metal chamber to stand (*i.e.*, without water cooling). The difficulty of getting a lining for such a combustion chamber that will stand the erosive powers of high-velocity, hot gases is obvious.

(2) The Furnace for the Mixed-fluid (Steam-and-Gas) Constant-pressure Gas Turbine.

The type of combustion chamber for this type of turbine becomes much simplified, owing to two reasons: first, the pressure in the chamber is that of the atmosphere (*vide* pp. 46 and 70); secondly, it is permissible to enclose the furnace completely with a water jacket, as the quantity of steam desired is the maximum quantity that can be produced for the initial temperature used. The type of combustion chamber for this turbine is shown in Fig. 41, and the arrangement by which the jacket can be made either for high-pressure or low-pressure steam exhibited. This is simply brought about by the opening of the valve K, and by-passing the steam through M into the furnace. The size of the furnace permits of lining the same with firebrick for such of its surface as is desired, and the large sectional area allows of comparatively low velocities of the gases through the furnace and the subsequent reduction of erosion effects to a minimum. If it be desired to burn the fuel in the furnace above the atmospheric pressure, it is at the disposal of the user so to do; the result of such procedure would only render the pressure difference between each side of the wall less than in the case of the high-pressure steam wheel and the low-pressure, mixed-fluid turbine.

It will be seen that in this type of combustion chamber the heat losses due to conduction and radiation are at a minimum. The temperature of the outside of the furnace walls cannot be greater than that of steam at 150 lbs. pressure, assuming that to be the pressure of the steam on the auxiliary steam wheel. This temperature is 180° C. (approximately). Taking the temperature of the surrounding atmosphere at 20° C., we have the temperature difference equals 160° C. In the case of the

Armengaud-Lemale combustion chamber the temperature of the outside of the walls would probably be about 500° C. (to put it at a low figure). This would mean a temperature difference of 480° C. It is difficult to estimate the proportion of heat loss due to radiation and that due to conduction. Supposing, however, we assume quite arbitrary values. Let the heat loss due to conduction in the case of the mixed-fluid turbine be ·03 H and that due to radiation ·005 H. In the Armengaud combustion chamber the loss due to conduction will be increased to 3 (·03 H), and the radiation loss to 3^4 (·005 H) = 81 (·005 H) = ·405 H. The loss has thus been increased from 3·5 per cent. to 49·5 per cent., or by the factor 14·1. The effect of an increase in the outside temperature of the walls of the combustion chamber is very apparent.

(3) The Combustion Chamber for the Closed-Chamber Explosion Turbine.

In the closed-chamber explosion turbine the time during which the chamber is exposed to the maximum temperature is very brief; in the constant-pressure, single-fluid turbine it is continuously exposed to this temperature. The loss in the former case, therefore, must be less than the loss in the latter for the same initial temperature. It is, however, in this case not practicable to line the chamber with non-conducting material, owing to the shock of explosion. The chief feature in this form of chamber is the valves. The combustion chamber of the Holzwarth turbine is shown in Fig. 67. The nozzle valve is held closed by a spring of such strength as to allow the valve to open immediately the maximum pressure is reached. By an ingenious system

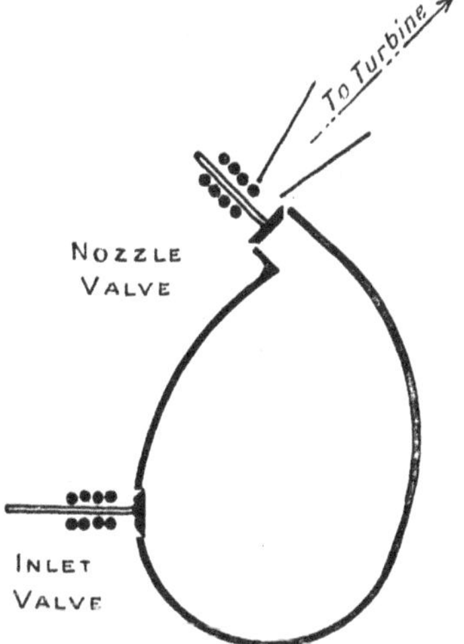

Fig. 67.—Explosion Chamber. Holzwarth Turbine.

of "oil control" (*vide* Chap. XI.) the valve is held open until the gases have entirely left the chamber. It is possible, of course, to water-jacket the explosion chambers here, as in the case of the constant-pressure gas turbine, but it would be uneconomical so to do, as here there is no large proportion of negative work, the evil effects of which the addition of steam would tend to minimise.

(4) The Combustion Chamber of the Karavodine Turbine.

The combustion chamber of the Karavodine type of gas turbine is, like the turbine itself, the most simple of all. The

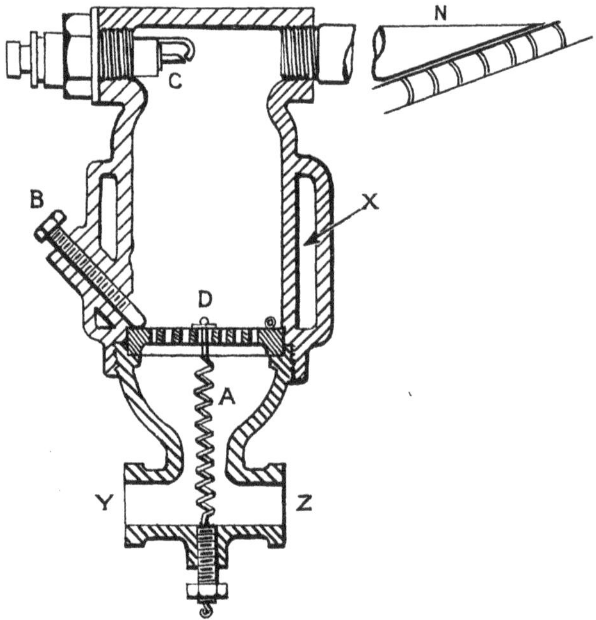

FIG. 68.—Karavodine "Explosion" Chamber.

form of chamber used by Karavodine with his experimental turbine is shown in Fig. 68. The only valve is the suction valve D, held down by a light spring, and opening automatically by the suction produced by the momentum of the exit gases. This type of combustion chamber is so simple that there is really little to be said about it; the type, having a momentum pipe added and coiled for compactness, is shown in Fig. 56

(*vide supra*, p. 129), and a regenerative type is here shown in Fig. 69.

(B) THE TURBINE WHEEL : WATER-COOLING.

Whether it is necessary to water-cool the turbine wheel in gas-turbine practice is rather a moot point. Enthusiasts on the non-cooling side, like Mr. R. M. Neilson, speak of using a temperature on the blades of 700° C., without any cooling device for the blades or wheel, which, as iron is a just visible dull red at 526° C., is, at least, sanguine. On the other side of the question, Herr Holzwarth considers it impossible to run a turbine permanently at a temperature of more than 450° C. (presumably with water cooling or without, as the assertion is stated generally).

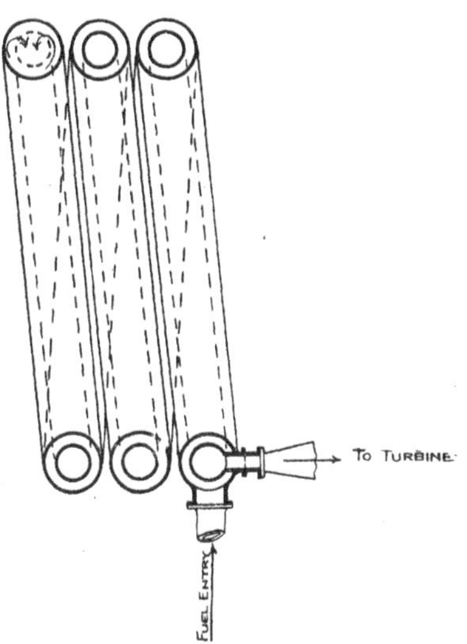

Fig. 69.—Karavodine "Explosion" Chamber with Momentum Pipe, Regenerative.

In the experimental turbine of Armengaud and Lemale the turbine wheel was a double wheel of the Curtis type, water cooled throughout, even the blades themselves being constructed with channels for the passage of the water. The difference in specific gravity between the hot and the cold water, together with the centrifugal action of the revolving wheel, was said to cause an automatic circulation. The great difficulty involved with the use of high temperatures on turbine blades is the erosion, or the melting away, of the fine edges of the blades. Unless these edges are tapered there is, of course, an undue loss in efficiency of the wheel due to the "shock" of the entering fluid on the blunt edges of the blades. With the ordinary type of Parson blade the fine edge is unavoidable. It is, however, possible to construct a wheel in which the sharp tapering away of the blade is avoided. If a wheel of section, as shown in Fig. 70, is milled out in the

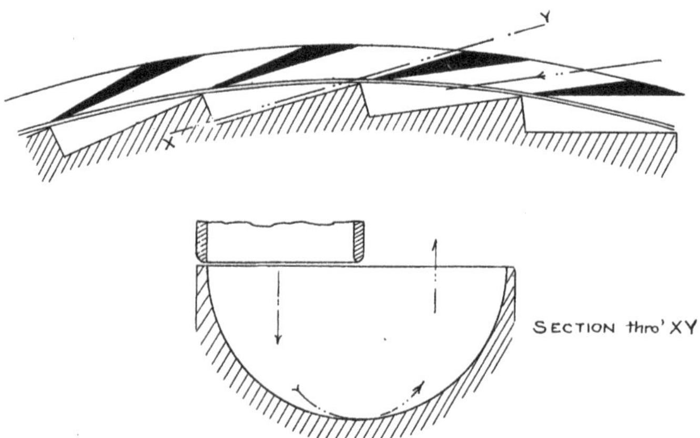

FIG. 70.—Milled Turbine Wheel.

manner there shown a turbine wheel, approximating to the Stumpf type, is the result. It will be noticed, however, that

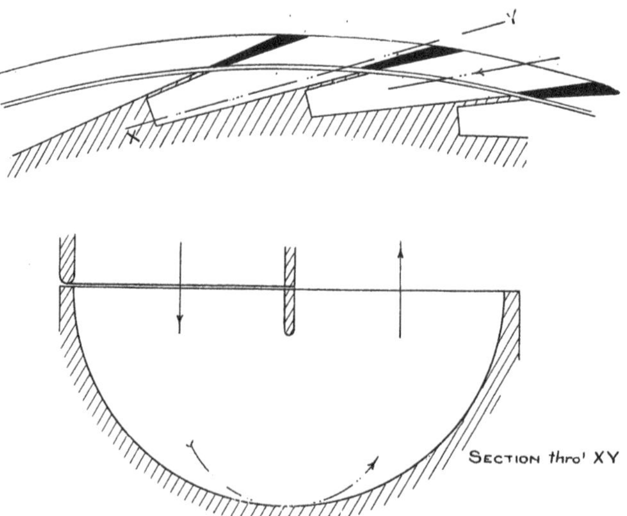

FIG. 71.—Stumpf Turbine Wheel.

whereas the Stumpf wheel (Fig. 71) is so milled out as to leave a dividing blade between successive buckets, here there is no such thin partition. The blade edges here are no longer of a finely tapering nature, the angle at the top of the blades being 90°. The angle in the case of the ordinary Parson blade is about 15°. It is to be expected, however, that this form of wheel should give a less efficiency than the true Stumpf type, owing to spilling through the absence of the blade partitions. But even with blade partitions the wheel of this type would be likely to endure higher temperatures than the Parson type of slotted blade, owing to the fact that with the

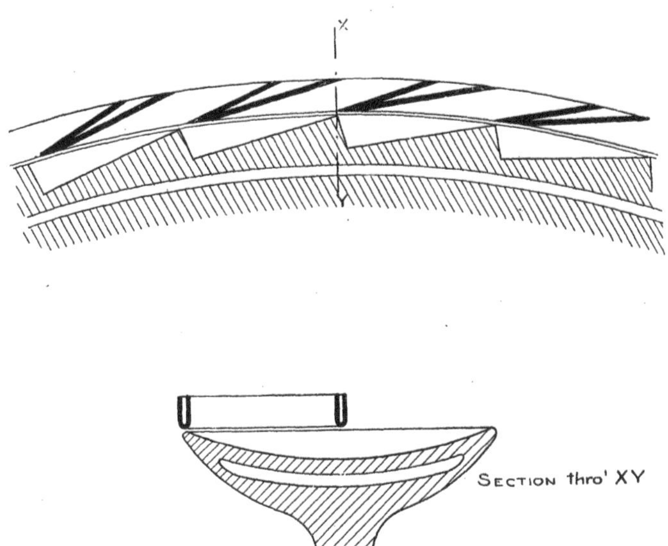

FIG. 72.—Milled Turbine Wheel with Water Cooling.

Stumpf wheel the blading is turned out of the solid. Perhaps this advantage hardly applies where, as in the case of Messrs. Escher, Wyss, the blades are *cast* solid on to the rim. "Double" Stumpf wheels have been constructed in which the fluid is reversed and led back upon the same wheel, but it is doubtful whether such are satisfactory. The cooling of the simplified Stumpf wheel is easy and efficient (*vide* Fig. 72). Owing to the unavoidable angle of inclination, the Stumpf wheel is less "efficient" than the ordinary De Laval type. As the temperature throughout the turbine is the same as that upon the blades (700° Neilson, 450° Holzwarth, 500° as taken here), it may be necessary to water-jacket the turbine casing. In the mixed-fluid turbine all waste heat has to go to the raising of steam, and there is no difficulty in water-jacketing every inch of the machine. In single-fluid turbines the loss of heat in the water jacket is not recoverable. Referring to the question of radiation (*vide infra*, p. 157), the loss in this case due to that phenomenon would be only about 3 per cent., assuming the temperature of the outside of the turbine casing to be at a temperature of 300° C. (same arbitrary values assumed). It is essential that the bearings be thoroughly water-cooled in all cases.

(C) THE REGENERATOR AND CONDENSER: THE SULPHURIC ACID QUESTION.

The use of the regenerator of the Stirling air engine type, in which the fluid first passes over some metal conductor, thereby parting with some of its heat, and then a fresh supply of fluid enters the regenerator, thereby taking up from the metal the heat that the outgoing gases have rejected, is not practicable in the case of the gas turbine (unless it be conceivable in the case of the explosion turbine). The flow of fluid must be continuous through the regenerator, and the heat has, therefore, to be taken up through metal walls. This, between air and air, is a wasteful process, and conduction of heat can only be brought about by large areas of heating surface and a high velocity of the issuing gases. For the mixed-fluid turbine the regenerator, of course, is little other than a locomotive boiler minus the furnace.

The condenser is of the orthodox form.

In this connection, however, arises the objection which has been most often urged against the mixed-fluid type of gas turbine.

Ordinary producer gas, or town gas, even when well scrubbed, almost invariably contains small quantities of sulphur. These cause considerable annoyance to the users of gas engines. The sulphur contained in the fuel takes, after burning, the form of sulphur dioxide, SO_2. When sulphur dioxide comes into contact with water an acid is formed, viz., sulphurous acid, in accord with the formula—

$$H_2O + SO_2 = H_2SO_3.$$

When a solution of sulphur dioxide (*i.e.*, sulphurous acid) is kept for any length of time in the presence of free oxygen, one molecule of oxygen is taken up, sulphuric acid being formed—

$$2H_2SO_3 + O_2 = 2H_2SO_4.$$

This action, however, is slow.

Now it is argued that the presence of steam in the waste gases from the mixed-fluid turbine, forming water in the condenser, will combine with the sulphur dioxide that may be present in the gas, to form, first, sulphurous acid, and, later, sulphuric acid ; and that the presence of this strong acid will be highly injurious to the metal surfaces to which it is exposed. That the danger is a real one there can be little doubt ; but one or two points may be noticed as to the true relevancy of the matter. It is doubtful, in the first place, whether sulphuric acid is actually formed in any perceptible amount. The oxidation of sulphurous acid is comparatively slow, and the atmosphere in which the reaction takes place contains a considerable preponderance of the inert gases, nitrogen and carbon dioxide. It would not seem that the presence of small quantities of sulphurous acid is to be greatly feared.

Even, however, if sulphuric acid is actually produced in the condenser, there are two ways open to the experimenter whereby its evil effects may be held in check. The first of these is to coat the condenser and the tubes thereof with some acid-resisting enamel ; a recoating would, of course, be required from time to time, but the matter is not more ponderous than that of cleaning fire tubes in a boiler ; there is no particular reason to suppose that any of the acid would be carried up into the pump (assuming the turbine to be sub-atmospheric), though, of course, such is possible.

The second, and more sure, method is to introduce into the condenser some base which would "neutralise" the acid formed—

$$X''(OH)_2 + H_2SO_4 = X''(SO_4) + 2H_2O.$$

Such a base as lime would suffice, sulphate of lime being

formed. The return of the condensed water to the furnace would then be prohibited, but the thermal loss due to this would be very small indeed. It is still indeterminate whether sulphuric acid is likely to be formed in any quantity ; and even if this be so, whether it is not curable.

(D) GOVERNING.

The question of governing is, in the case of the gas turbine, an extremely difficult one. In the case of the explosion turbine the matter is comparatively simple, the control of the number of explosions made per unit of time, or of the number of nozzles in operation, being a matter of easy manipulation. In the continuous-combustion turbine, however, the problem does not allow of so satisfactory a solution. It is not here permissible to alter the size of the nozzle entry, as the speed of the pump would have to be correspondingly changed ; also such regulation is not delicate enough for speed control.

In the case of the steam-and-gas turbine described in Chapter IV., with the additional high-pressure steam wheel, the governing is especially simple and delicate. This can be effected altogether on the steam wheel by by-passing the steam directly into the furnace. The by-passing of the steam altogether would drop the overall efficiency from 34 per cent. to 28·5 per cent. A slight by-passing of the steam would, however, suffice for the governing, causing not more than 1 per cent. variation in the overall efficiency. The governing would at once be responsive and flexible.

With types of constant-pressure turbines *not* using steam expansion on separate wheels there are two methods by which the peripheral speed of the turbine can be controlled : one by the alteration of the initial temperature ; the other by the alteration of the expansion ratio—*i.e.*, by the alteration of the lower limit of pressure, the " breaking of the vacuum."

The effectiveness of the two methods may be analysed as follows :—

The work done by 1 lb. of air in expanding from a pressure p_1 a pressure p_2, adiabatically, is—

$$W = \frac{\gamma}{\gamma - 1} v_2 p_2 \left(R^{\frac{\gamma - 1}{\gamma}} - 1\right)$$

where $R = \frac{p_1}{p_2}$; cf. T_2 = temperature corresponding to p_2, and $p_2 = p_0$.

$$\therefore W = \frac{\gamma}{\gamma - 1} v_0 p_0 \cdot \frac{T_2}{T_0} \left(R^{\frac{\gamma-1}{\gamma}} - 1\right)$$

Now $\frac{\gamma}{\gamma - 1} \cdot \frac{v_0 p_0}{T_0} \cdot \frac{1}{J} = C_p = \cdot 28$.

$$\therefore W = \cdot 25 \, T_1 / R^{\frac{\gamma-1}{\gamma}} \left(R^{\frac{\gamma-1}{\gamma}} - 1\right)$$

$$\therefore W = \cdot 25 \, T_1 \left(1 - 1/R^{\frac{\gamma-1}{\gamma}}\right)$$

Now rate of change of W with R—

$$\frac{dW}{dR} = \cdot 25 \, T_1 n R^{-(n+1)}$$

when $n = \frac{\gamma - 1}{\gamma} = \cdot 275$.

The rate of change of W with T—

$$\frac{dW}{dT} = \cdot 25 \, T_1 (1 - R^{-n})$$

Therefore—

$$\frac{dW}{dR} : \frac{dW}{dT} :: 4 \cdot 8500 : \cdot 1175.$$

Or governing by "breaking the vacuum" is forty-two times more effective than governing by lowering the initial temperature, keeping, of course, the right units for R and T, namely, one expansion compared to 1°. This means that 1 inch of water change in the vacuum is as effective as a temperature change of 9° C. Also, it should be borne in mind that the method of governing by breaking the vacuum in the condenser operates much more swiftly than any system of temperature or, perhaps, even nozzle, control.

(E) THE PUMP.

The ultimate success of the constant pressure gas turbine as an economical heat engine depends very largely on the pump. In the explosion turbine of the Lenoir type, where no initial compression of the charge is undertaken, the pump, of course, ceases to be a controlling factor in the efficiency of the machine.

As, however, this particular class of explosion turbine has shown but poor promise of success in fuel consumption, the compressor still remains more than a mere pawn in the practical

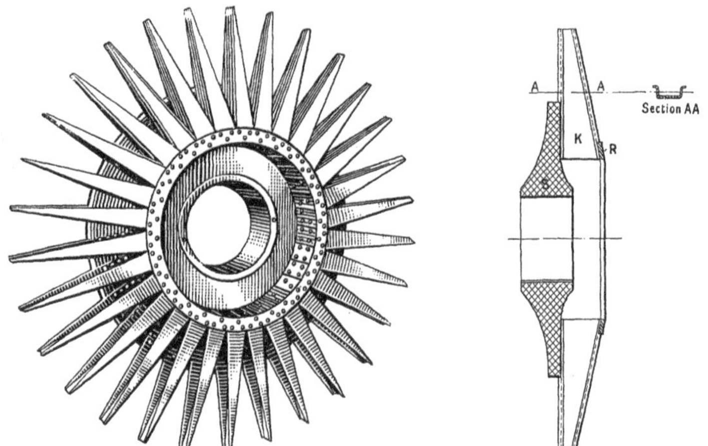

FIGS. 73 and 74.—Single Rotor for Zoelly Compressor.

solution of the problem. Even in the explosion turbine of Herr Holzwarth the charge is initially compressed by a rotary blower to 12 lb. or so above the atmosphere. The importance of the compressor here, of course, is not of the same magnitude as it is in the steam-and-air turbine. The use of piston com-

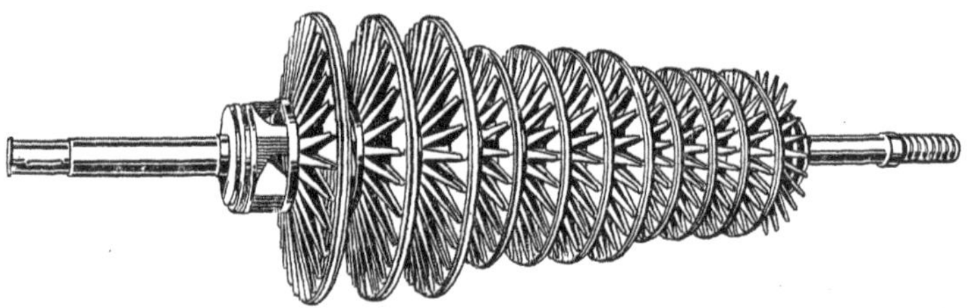

FIG. 75.—Complete Rotor of Zoelly Compressor for seven Compressions.

pressors is almost entirely prohibitive on account of the extreme bulkiness of such machines dealing with the quantities of air required in gas turbines.

The rotary compressor is, in its essence, extremely simple. It consists of a rotor, bearing radial—or sometimes curved—

blades, which deliver the air through an expanding nozzle—diffuser—to a series of guide blades, which in a multiple machine conduct the air into the eye of the next wheel. A sufficient number of these rotors are mounted upon a continuous shaft, the exhaust from one delivering into the eye or inlet of the other.

There is but small divergence in the form of rotor and pump

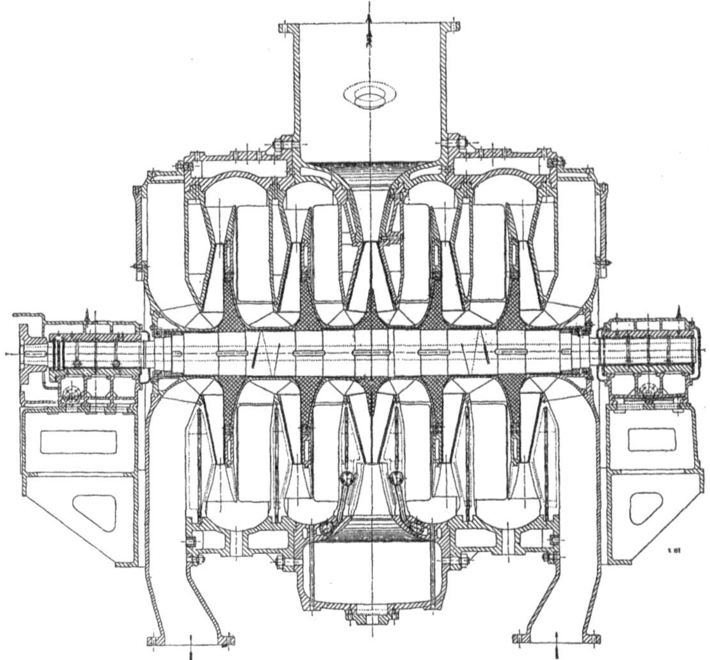

FIG. 76.—Zoelly Rotary Compressor.

chamber employed. In the Zoelly pump the rotor blades are radial, and cupped into the form of a rectangular channel; they are only shrouded for about half their length from the inner margin. In the Rateau compressor the blades are not simply radial but curved, commencing radially on the inner edge and finishing more or less tangentially at the periphery. The blades are shrouded for the whole of their length.

In the Zoelly form of rotary compressor guide blades are introduced at the eye of the rotor and the expanded diffuser into which the rotor delivers is also fitted with blades. In the Rateau pump the rotor delivers into a "free vortex," guide blades being employed to conduct the air towards the centre of the ensuing wheel.

An illustration of the rotor for a Zoelly compressor (by

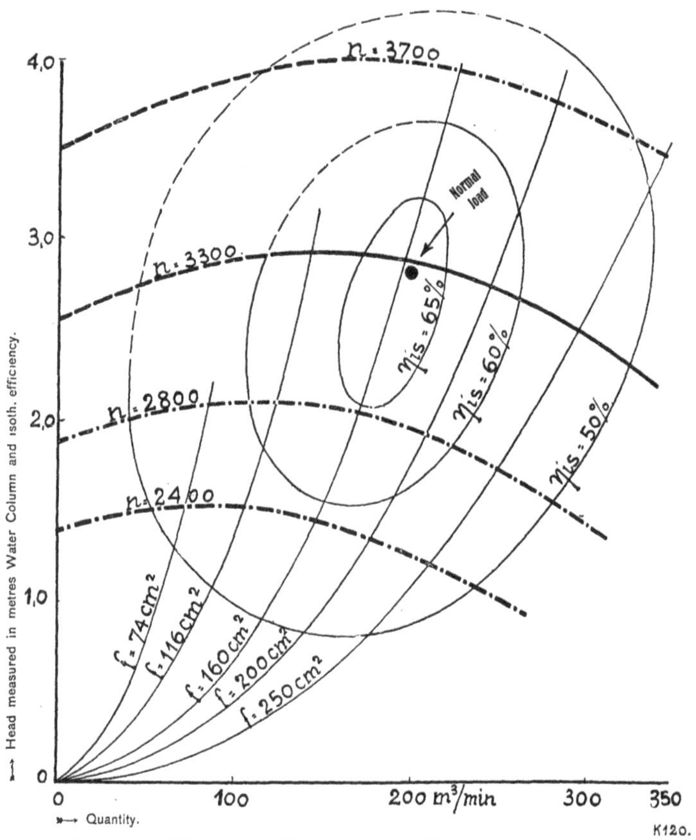

Fig. 77.—Efficiency Diagram for Zoelly Compressor.

Messrs. Escher Wyss) is shown in Figs. 73 and 74, and in Fig. 75 is shown the whole rotor for a compression of one to seven atmospheres absolute. A section of a double "balanced" compressor by the same firm is shown in Fig. 76. The system of water-jacketing is clearly shown. A diagram giving the

FIG. 79.—Rateau Multiple High-Pressure Compressor, used in Conjunction with the Armengaud-Lemale Gas Turbine.

performance of the Zoelly pump is shown in Fig. 77 ; it will be seen that about 65 per cent. seemed to be the maximum efficiency obtained.

The construction of the Rateau rotary compressor is shown in Fig. 78. It will be seen that it differs from the Zoelly types in the absence of the "diffuser" blades, the air being delivered into a free vortex at S. It will be seen also that the rotor blade is completely shrouded on both sides. Fig. 79 shows a multiple high pressure, Rateau compressor. The one here reproduced was that used in the Armengaud-Lemale experiments with the constant-pressure gas turbine. The pump in question pumped from atmospheric pressure to 112 lbs. per square inch absolute (R = 7·6). It gave an efficiency of 65 per cent.

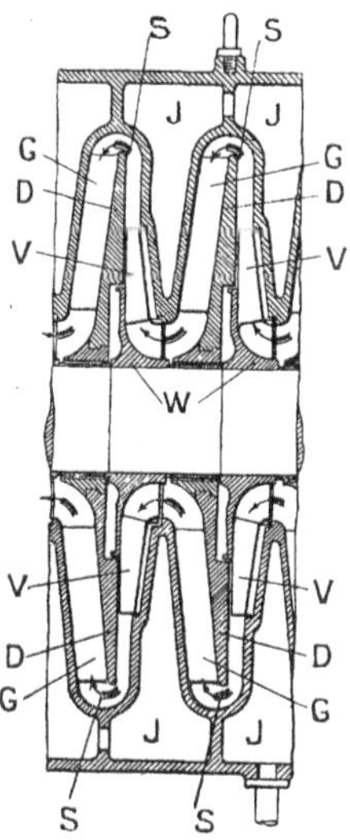

FIG. 78.—Section through Rateau Compressor.

Owing to the density of air being low, the pressure produced by a given peripheral speed is necessarily small, since the pressure P, produced by a given angular velocity, is given by the formula—

$$P = \omega^2 \int_{r_2}^{r_1} \rho r \,.\, dr,$$

Where ω = angular velocity, ρ = density of fluid, r = radius of rotor.

With a non-expansible fluid, such as water, the density remains constant throughout the rotor, and—

$$P = \tfrac{1}{2} \rho \omega^2 (r_1^2 - r_2^2)$$

With air the value of ρ changes with that of r.

Professor Rateau gives the following formula for the rise in pressure produced by a single rotor, starting from atmospheric pressure—

$$h = \frac{\mu \rho u^2}{g}$$

h being pressure in kilogrammes per square metre.
ρ density in kilogrammes per cubic metre.

u the peripheral velocity of rotor in metres per second.
g the acceleration of gravity in metres per second per second.
μ a coefficient, varying with the form of the rotor blades—with the Rateau compressors it lies between 0·50 and 0·55.

This gives the relation between peripheral speed and ratio of compression, which is independent of the actual value of the initial pressure, since the work done for a given ratio of compression is constant.

If R is the ratio of the final to the initial pressures—or the ratio of compression—then—

$$R = 1 + 6 \times 10^{-7} v^2.$$

Where v = peripheral velocity of rotor in feet per second.

Or—

$$R = 1 + 1{\cdot}64 \times 10^{-9} \times N^2 d^2.$$

Where N = number of revolutions of rotor per minute; and d = outside diameter of rotor in feet.

Thus, with a peripheral velocity of 600 feet per second a compression ratio is obtained equal to 1·216. The increase of compression with the number of elements employed increases in geometrical progression. That is to say, if r is the ratio of compression for one element and if R is the final compression ratio desired, then—

$$r^n = R$$

where n is the number of elements.

Thus, if six compressions are required, and if the peripheral velocity is 600 feet per second, at least nine elements will be necessary.

For low-pressure compressors it is usual to make them in duplicate; the air being taken in at both ends and delivered at the centre, or drawn in at the centre and out at the ends.

In cases where greater ratios of compression are required—that is, compression ratios of six to ten, such as are necessitated in the gas turbine—the number of elements required becomes of such magnitude that it is no longer permissible to balance the machine by the simple process of duplication. In these cases the compressor is divided into two or more groups of elements separated by bearings, the air being made to traverse the "cylinder" (*vide* note, *infra*) in opposite directions, so that a balance may be obtained.

With rotary compressors in which the ratio of compression is small, and the number of elements consequently few, high efficiencies have been obtained—80 per cent. being by no means an uncommon figure, while values approaching even 90 per cent. have been reached. With greater ratios of compression the efficiency, unfortunately, is considerably lower. The rotary pump used in the Armengaud-Lemale experiments gave an efficiency of about 65 per cent. This result has, however, been since improved upon, and efficiencies of 70 per cent. have been achieved with compressors delivering air up to 100 lbs. pressure per square inch.

Apart from the rise in temperature caused by insufficient cooling and the consequent departure from the conditions of

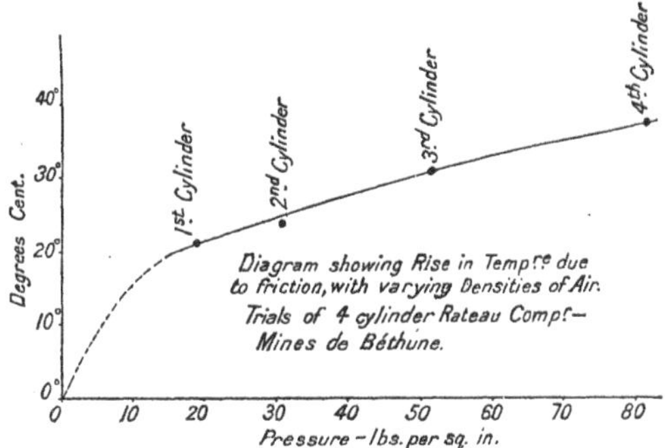

Fig. 80.—Pressure-Temperature Diagram for Rateau Compressor.

isothermal compression, the only losses that remain to be accounted for are leakage, friction in bearings, and the air friction of the revolving parts. Of these the last item is the most important. The work taken up in air friction is directly dependent upon the density of the air in which the rotor revolves. Fig. 80 is a diagram showing a connection between the density of the air and the work lost in friction. It is only indicative in nature, and is based upon the trials of a four-cylindered Rateau compressor at the Mines de Béthune. In this instance the air was not cooled between each element, but only between each group of elements—that is, between each "cylinder."[1] The temperature at the entry and at the

[1] The above data is from a paper by Professor Rateau on high-pressure, rotary air compressors. It is difficult to find an English

exit of each group of elements was ascertained. This was found to be in excess of the temperature theoretically due to the work done in the adiabatic compression. The difference in temperature between the actual value recorded and that theoretically determined can only have been produced by the work done in the friction of the rotors against the air in which they revolved. If, therefore, this excess temperature rise for each "cylinder" is plotted against the mean pressure in each "cylinder," some indication will be afforded as to the increase in frictional losses with the change in density of the medium.

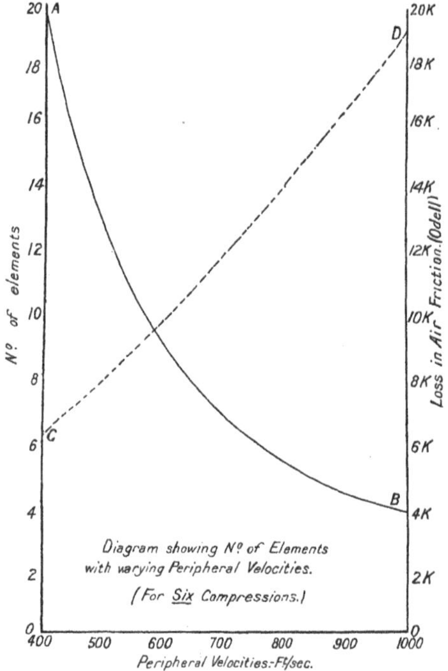

FIG. 81.—Graph of Rateau and Odell Formulæ.

When the density of the medium in which the rotor revolves is zero, the frictional resistance must of necessity vanish. The curve of the locus of the temperature pressure values for the four "cylinders" must therefore pass through the origin. This is indicated by the dotted portion of the graph; its form is, of course, indeterminate.

The above mentioned data afford no quantitative evaluation of the work lost in friction; they merely bear out the formulæ for air friction with revolving discs of Odell, Lewecki, Stodola, etc., in which the friction is directly dependent upon the density of the surrounding air.

word to denote the "group of elements" of a multiple compressor. The word used by Professor Rateau is "corps"; a direct translation would be meaningless. I have substituted the word "cylinder" as the nearest equivalent.

The moral of the matter in gas turbine practice is clear. In cases where no more than six or seven expansions are employed in the turbine it is obviously more efficient to perform this expansion below the atmosphere than above it. That is, to use the rotary pump not as a compressor, but as an exhauster. With six expansions below the atmosphere, for example, the mean density of the air is only 8·75 lbs. per square inch. With six compressions above, the mean pressure is 52·5 lbs.—a density six times as great. According to Odell's experiments the work lost in air friction is directly dependent on the density of the medium in which the disc revolves. In the latter case, therefore, the frictional losses would be increased to six times their former value. Referring to the peripheral speed necessary to obtain a certain ratio of compression, it is obvious that the number of elements may be very considerably reduced (under any fixed total expansion ratio) by the increase of the peripheral speed. In Fig. 81 is shown the variations in the number of elements, for six compressions, with variable peripheral velocities of from 400 to 1,000 feet per second (curve AB). It is naturally desirable that as high a peripheral velocity as is practicable should be used, as by so doing the bulk of the machine is very considerably reduced. The objection to high rotational speeds is an economic one. According to Odell's experiments with the air friction of smooth discs (approximating to the case of the Rateau shrouded wheel), revolving in free air, the work lost in air friction—

$$L = 0{\cdot}02295 \,.\, k \,.\, D^{2{\cdot}5} \, (u/100)^3 \,.\, \rho.$$

Where D is the diameter of the disc in feet.

u is the peripheral velocity in ft./sec.

ρ is the density of the medium.

k is a constant ; approximate value, 3·14.

It will be thus seen that the frictional loss depends upon the *cube* of the peripheral speed. It is also directly proportional to the number of elements (for any given output). Therefore, in the diagram above mentioned (Fig. 81) if these losses are plotted against the varying velocities, taking into account the change in the number of elements, we have the equation, loss due to friction—

$$L = K \,.\, n \,.\, u^3.$$

Where n is the number of elements.

u the peripheral velocity.

K a constant, depending upon the size of the rotor and the conditions of working.

This equation is shown in Fig. 81 by the broken line CD. The value of this constant, K, can be determined, approximately, on certain assumptions, from purely experimental data.

In Fig. 82 is shown the performance curves for a Rateau turbo-compressor of 328 H.P. pumping to 50 lbs. per square

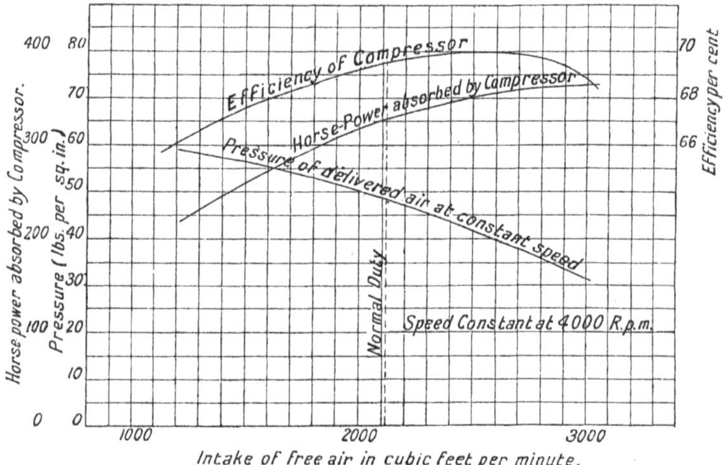

FIG. 82.—Performance Curves for Rateau Turbo-Compressor.

inch, at a constant speed of 4,000 r.p.m., by Messrs. Brown, Boveri & Co. It will be seen from the curves in question that, when delivering to a pressure of 60 lbs. abs., the pump gave an efficiency of 66 per cent.; and when pumping to a pressure of 40 lbs. abs., an efficiency of 70 per cent. The peripheral speed is constant; the surface of the rotating body is constant; the only condition affecting the value of the loss due to air friction is the drop in pressure and the subsequent change in the mean density of the medium. It is assumed that this change does not affect any other loss but that of the air friction.

Consider the theoretical output to be 100 units of work—

We have total loss, Z_1, at mean density, 37·5 lbs. abs. = 34.

total loss, Z_2, at mean density, 27·5 lbs. abs. = 30.

Let all losses other than that due to air friction be represented by λ. Let loss due to air friction at maximum pressure, 60 lbs. ($\rho = 37\cdot5$ lbs./sq. in., abs.) be denoted by L_1; that at maximum pressure, 40 lbs. ($\rho = 27\cdot5$ lbs./sq. in. abs.), be denoted by L_2. Then we have the equations—

$$Z_1 = L_1 + \lambda$$
$$Z_2 = L_2 + \lambda.$$

But

$$L_1 = C' . \rho = C' . 37\cdot5$$
$$L_2 = C' . \rho = C' . 27\cdot5,$$

where C' is the constant when ρ is given in terms of lbs. per square inch absolute.

We have, therefore—

$$34 = C' . 37\cdot5 + \lambda$$
$$30 = C' . 27\cdot5 + \lambda$$
$$4 = C' . 10$$

Therefore

$$C' = \cdot4.$$

And the loss due to friction, $L = \cdot4\rho$, where ρ is reckoned in lbs. per square inch absolute.

Thus, when the compressor is pumping from the atmosphere to 60 lbs. absolute, the loss due to friction—

$$L = 37\cdot5 \times \cdot4 = 15 \text{ per cent.}$$

and the losses due to other causes 19 per cent.

Consider the same machine (λ remaining constant) to be used for the sub-atmospheric air turbine, compressing from 1·5 lbs. to 15 lbs. abs. The mean density in this case is 8·25 lbs. abs. Therefore—

$$L = 8\cdot25 \times \cdot4 = 3\cdot3 \text{ per cent.}$$

The loss due to causes other than disc friction is still 19 per cent., and therefore the total loss in this case is 22·3 per cent.

This means that the thermodynamic efficiency of the rotary compressor working entirely below the atmosphere, as proposed in the case of the mixed-fluid gas turbine described in Chapter III., is 77·7 per cent., whereas the thermodynamic efficiency of the pump compressing from the atmosphere to 60 lbs. abs. is only 66 per cent.; or, in the case of the same ratio of expansion above, to that below the atmosphere (*i.e.*, 10), 48 per cent.

This shows very clearly the immense advantage of sub-atmospheric expansion for the constant-pressure gas turbine mixed-fluid or otherwise.

CHAPTER VIII. PRACTICAL LIMITATIONS

THE question that is most pregnant in the gas turbine problem is the fixing of what are the limiting conditions under which the machine may satisfactorily operate. The pressing fundamental questions that have to be answered before any result can be calculated, are: (1) What is the maximum temperature that can be used in the cycle? (2) What is the maximum peripheral speed at which the turbine wheel may be run? (3) What is the maximum pressure that can be employed; what is the minimum pressure that can be employed; and, from these, what is the maximum expansion ratio that can be used?

Thus we have the following limiting conditions to fix—

(*a*) Maximum temperature before expansion, T_1.
(*b*) Maximum temperature after expansion, T_2.
(*c*) Maximum peripheral velocity of wheel, u.
(*d*) Maximum pressure in cycle, p_1.
(*e*) Minimum pressure in cycle, p_2.
(*f*) Maximum expansion ratio, R.

Now these conditions are not independently variable, but there is one primary condition from which all the others emanate; this is the maximum temperature that can be used after expansion—the maximum temperature, T_2, that can be used on the turbine blades. Having settled this value, T_2, we can then find the maximum temperature, T_1. There is, of course, a limiting value of T_1, independent of any other factor, the temperature, in short, which the material of the combustion chamber can no longer stand; this value, however, is outside the range of practice, and need not be considered here. The limiting value of T_1 is really fixed by the quantity of work that the turbine can absorb, this work being proportional to $(T_1 - T_2)$. If u be the peripheral velocity of the turbine wheel, we have—

$$T_1 = \phi\ (u,\ T_2),$$

where ϕ denotes "function of." (u is regarded as equal to

$n \cdot u'$, where n is the number of rings of blades on the initial turbine wheel and u' is the actual value of the peripheral velocity; u really represents the "work-absorbing" power of the wheel.)

The value of u is given by—

$$u = \phi_1 (u', n)$$

where u' and n have to be fixed as independent constants.

The maximum and minimum pressures in the cycle p_1 and p_2 have absolute limiting values, the first being dependent on the maximum pressure to which the pump used can pump (or in the case of the explosion turbine the maximum pressure which the explosion vessel can stand); the second dependent on the lowest vacuum that can be realised with the type of exhauster used. Both p_1 and p_2, however, have limiting values within these, dependent on the expansion ratio that it is allowable to employ. This is expressed by—

$$p_1 = \phi_2 (R, p_2)$$
$$p_2 = \phi_3 (R, p_1).$$

Lastly, we have relation—

$$R = \phi_4 (u, T_2).$$

We have now to consider individually the values of the above-mentioned limits under conditions of normal practice.

In the preceding chapters, it will be remembered, T_2 was taken constant at 500° C.; u was taken constant at 850 ft./sec.; p_2 was taken at 1·5 lbs.; p_1 (about) 90 lbs. abs.; R, as an independent value, once at 20 (p. 38), once at 10 (p. 56). The orthodoxy of these assumed values has now to be discussed.

I. THE TEMPERATURE PERMISSIBLE UPON THE TURBINE BLADES.

In the last chapter it was pointed out that Mr. R. M. Neilson, in his paper before the Institution of Mechanical Engineers,[1] suggested using a temperature of 700° C. (without water-

[1] October 21st, 1904.

cooling), and a temperature of 1,500° C., or even 2,000° C. (with water-cooling).

In the same year (1904) as Mr. Neilson's paper, appeared the first notice of the experimental work of Armengaud and Lemale (A. Barbezat, *Schweizerishe Bauzeitung*, August 27th). The wheel of their turbine, which was water-cooled, ran, in this case, at a temperature of 400° C., a considerable drop from the temperatures mentioned by Mr. Neilson of 1,500° to 2,000° C. Four years later, in the winter of 1908, were begun the experiments made by Herr Holzwarth, of Mannheim, upon the explosion turbine. In October, 1911, these experiments reached a head in the construction of a "thousand horse-power" turbine. In the tests published it appeared that the value of T_2 varied between 457° C. and 947° C. In the *Journal fur Gasbeleuchtung* (Munich) for September of last year Herr Holzwarth asserts: "that in my experience a turbine cannot be worked permanently at more than 450° C." This is the state of affairs at the present moment as far as the author knows. It must be remembered, however, that the wheel of the Holzwarth turbine was not water-cooled, and it may be, perhaps, that his remarks refer only to non-cooled rotors.

The evidence provided, it must be admitted, is not overwhelming. At 526° C. iron exhibits a dull red heat that is just visible. Without any system of water-cooling it may, perhaps, be assumed that it would be possible to use a temperature of 500° C. upon blades specially composed of some resistant alloy. This we may regard as the maximum limit for T_2 with non-water-cooled turbine wheels, though it is only fair to admit that Herr Holzwarth, who certainly has a better right to speak on such a subject than any other living man, will not allow a value exceeding 450° C.

With an efficient system of water-cooling it might be possible to raise this value of T_2 from 500° C. up to (say) 700° C. It may be taken that 200° C. represents the increase in temperature allowable by the introduction of artificial cooling processes. This may be regarded—until we have some further experimental data to hand—as the maximum possible temperature limit for water-cooled blades—a sufficient falling off from the figures proposed by Mr. Neilson in 1904.

It may be assumed, without too great a risk of being wrong,

N 2

that 400° C. without water-cooling, and 500° C. with water-cooling, are eminently sane values for T_2, in actual present-day practice. We have, therefore—

	By common practice.	Maximum value.
T_2, without water-cooling . . .	400°	500° C.
T_2, with water-cooling . . .	500°	700° C.

It is by no means denied that higher values than these may be obtained by use of cooling devices as yet unconceived, or with blade material, as yet undiscovered. But to base constants or limits on the visionary results of future experiments seems neither logical nor capable of being of much use to anyone.

II. THE PERIPHERAL VELOCITY PERMISSIBLE FOR TURBINE WHEELS.

Unlike the case of the maximum temperature that it may be possible to use on turbine blades, we here have plenty of experimental evidence and, what is much more, evidence of common working practice, as to the maximum speed at which turbines may be run. Turbines that are designed for special purposes, *i.e.*, ship propulsion, etc., require specially low peripheral speeds. It is not necessary that, at the present juncture in the evolution of the gas turbine, we should regard these. What we more particularly have to deal with is the gas turbine unit driving dynamos or alternators. The speed at which such may be run is (within wide limits) arbitrary. What we are more immediately concerned with is the maximum peripheral speed allowable in ordinary steam turbine practice for the turbine rotors.

It must be remembered that the drop in temperature of the working fluid from T_1 to T_2 must take place before entering the first wheel, and that all the kinetic energy thus produced must be absorbed, as far as possible, upon the first wheel, whether or no any further expansion below T_2 on other wheels takes place. The species of turbine required, therefore, must be, essentially, of either the De Laval or the Curtis type, or an intermediate form between these two, that is, a one-wheel turbine with a double ring of blades. This last type was used in the experiments both of Armengaud-Lemale and Holzwarth.

De Laval has run his turbine wheels at over 1,300 ft./sec., but the diameter of the rotor was under 30 inches, and the necessary reduction of shaft velocity to that suitable to the driven unit

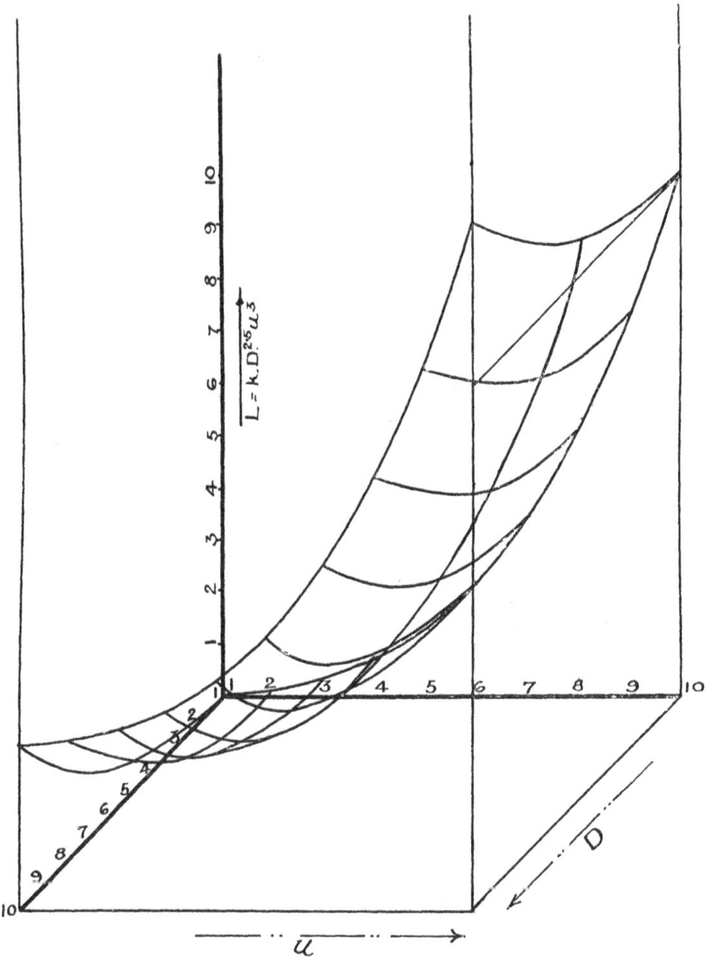

Fig. 83.—Three-Dimension Diagram, showing Variation of Frictional Loss of rotating disc with Diameter of Rotor and Peripheral Speed.

had to be effected by gearing. It must be borne in mind that all losses due to the friction of the rotor in the medium in which it revolves is proportional to the 2·5th power of the diameter of the rotor, the cube of the peripheral speed, and directly to the density of the medium (Odell). That is to say, we have loss due to friction—

$$L = k \cdot D^{2 \cdot 5} u^3.$$

Consider a particular instance. Let loss due to friction be λ, when D is 3 feet and u is 100 ft./sec. Let D be increased to 6 feet, and u to 600 ft./sec., we have—

$$L = 610 \lambda.$$

Or, if in the first instance the rotational frictional loss equalled ·05 per cent., in the second case it would equal 30·5 per cent. We have here a factor varying with two variables. The results of this variation are shown on the three-dimensional diagram (Fig. 83). The relative increase in the frictional loss L is obtained for any values of D and u by dropping perpendiculars on to the base surface.

We need not consider the question of geared gas turbines of the De Laval type ; any possible increase in work done, gained by the increase of the speed of the rotor, being more than counteracted by the loss in the gearing.

The highest peripheral speed used in practice with large-wheel, non-geared turbines was attained with the Stumpf turbine, which ran at speeds of 1,000 ft./sec. or over. An 8-foot wheel running at 2,400 r.p.m. would produce this peripheral velocity. This is, however, still an excessive figure. It would, perhaps, be more reasonable to take 900 ft./sec. as a maximum and 800 ft./sec. as a probable value for u in actual practice.

III. THE MAXIMUM VALUE OF THE EXPANSION RATIO.

The maximum expansion ratio allowable is determined by the amount of work that the first wheel of the turbine can absorb and the maximum temperature which that wheel can withstand, assuming, of course, that there is no further expansion below this temperature. This value of R will be greater for the constant-volume turbine than for the constant-pressure turbine, as the work done per unit weight of fluid is greater in the latter case than in the former. The determination of R for the two cycles is obtained in the following manner :—

(*a*) For the constant-pressure, single-fluid gas turbine—

$$W = C_p\,(T_1 - T_2)$$
$$= \frac{(n\,u)^2}{2g\,.\,J},$$

where u is the actual peripheral velocity of the wheel, and n represents the ratio $\frac{v}{u}$, v being the velocity of the working fluid ; n is dependent on the number of rings of blades on the wheel (*i.e.*, two usually) and the angle of entry and residual velocity of the waste gases leaving the turbine. With single Curtis wheel turbines n may be taken as equal to 5 (*vide* p. 31).

Now—

$$T_1 = T_2\,.\,R^{\frac{\gamma-1}{\gamma}}.$$

Therefore we have—

$$\frac{(nu)^2}{2g\,.\,J} = C_p\,(T_2\,.\,R^{\frac{\gamma-1}{\gamma}} - T_2),$$

$$\therefore\; R = \left[\frac{(nu)^2}{2g\,.\,J\,.\,C_p\,.\,T_2} - 1\right]^{\left(\frac{\gamma-1}{\gamma}\right)}.$$

We can thus get for limiting values of R, a maximum possible value when $T_2 = 500°$ C. (or 700° C.), and $u = 900$ ft./sec., and a more likely value when $T_2 = 400°$ C. (or 800° C.), and $u = 800$ ft./sec.

(*b*) With the constant-pressure, *mixed-fluid* turbine the *absolute* calculation of R is extremely difficult.

The value can be approximated by taking an assumed value for y, the quantity of steam added to the products of combustion and approximate values for τ_2 in the Rankine formula—

Thus we have—

$$\frac{(nu)^2}{2g\,.\,J} = \left\{ \left[C_p\,(T_2\,.\,R^{\frac{\gamma-1}{\gamma}} - T_2)\right] + y\left[(\tau_1 - \tau_2)\left(1 + \frac{L_1}{\tau_1}\right) + C_{p_s}\right.\right.$$
$$\left.\left.(T_2\,.\,R^{\frac{\gamma-1}{\gamma}} - \tau_1) - \tau_2\left(\log_e \frac{\tau_1}{\tau_2} + C_{p_s}\,.\,\log_e \frac{T_2\,.\,R^{\frac{\gamma-1}{\gamma}}}{\tau_1}\right)\right]\right\} \div \{1 + y\}.$$

The values of y, τ_2 (τ_1 being fixed) are, of course, dependent on R.

(*c*) For the constant-volume, single-fluid turbine. Here we have—

$$W = C_v (T_1 - T_0) - C_p (T_2 - T_0).$$

$$\therefore \frac{(nu)^2}{2g \,.\, J} = C_v \left\{ R \,.\, ^{\frac{\gamma-1}{\gamma}} T_2 - T_2 - \gamma T_2 + \gamma T_0 \right\}.$$

$$\therefore \; R = \left[1 + \gamma - \frac{1}{T_2} \left\{ \gamma \,.\, T_0 + \frac{(nu)^2}{2g \,.\, J \,.\, C_v} \right\} \right]^{\frac{\gamma}{\gamma-1}}.$$

This gives two values, as before, for R maximum, according to the maximum value chosen for T_2 and u.

IV. THE MAXIMUM AND MINIMUM PRESSURES POSSIBLE IN THE CYCLE.

The determination of the maximum pressure, p_1, is, in the constant-pressure gas turbine, concerned directly with the capabilities of the pump. With a piston pump there is, of course, no limit, theoretically, to which the gases may be compressed, as the compression takes place above the critical temperature of the gases in question. Both the size of the pump and the inefficiency consequent at high pressures, however, soon set a limit to which the rotary compressor can be used.

The Rateau compressor used in the Armengaud-Lemale experiments compressed the air from the atmospheric pressure to 112 lbs. per square inch absolute. At this compression it gave 65 per cent. efficiency. It, perhaps, may be safely said that 150 lbs. per square inch absolute represents the maximum value to which rotary compressors could pump with anything like a reasonable number of elements and a commendable efficiency. A value more coincident with common practice might be taken at 100 lbs.

The determination of the limiting value of the minimum pressure to which a rotary compressor could exhaust is extremely difficult, on account of the absence of any data as to the performance of these machines under sub-atmospheric conditions. That, both from the experimental data available (*vide* p. 174), and also from theoretical considerations (*vide* p. 173), the rotary compressor would be more efficient at low pressures there can be but little doubt. To place a limit to the terminal pressure at which air becomes so rarefied as to cease to follow common centrifugal laws is impossible. In the absence of all experimental evidence concerning the matter, the author has fixed the value of p_2 at 1·5 lbs. per square inch ; but there seems no adequate reason why vacua of 1 lb. or even ·5 lb. should not be obtained.

The determination of the maximum value of p_1 with the explosion turbine is, of course, determined by the maximum value of R, assuming that the terminal pressure is that of, or approximately that of (*vide infra*, p. 228), the atmosphere.

Two factors that do not exactly come under the head of "limiting conditions," but which determine the overall efficiencies of the various types of gas turbines, are the thermodynamic efficiencies of the turbine and rotary pump. These, of course, have to be taken at such values as these units give in practice. Let us consider the case of the turbine.

The best turbine thermodynamic efficiencies obtained with steam turbines approach something like 70 per cent. This figure has been adopted so far in the working out of the overall efficiencies. But two points have to be borne in mind. First, one of the chief sources of loss in the steam turbine, namely, the premature condensation of the steam and its presence in the blade channels, is eliminated in all gas turbines ; even in the mixed-fluid turbine the conditions of superheat admit of no such condensation. Secondly, in the sub-atmospheric turbine, the loss due to "disc" friction is greatly decreased compared to that in super-atmospheric turbines (this is one of the reasons that go to making low-pressure steam turbines much more efficient than high-pressure ones). On these grounds it is not altogether unreasonable to assume that, except in the case of the explosion turbine, a higher thermodynamic efficiency is likely to be obtained than that which is obtained at present in the steam turbine. This higher value may be set at 75 per cent.

With regard to the rotary pump, in Rateau compressors efficiencies have been obtained of 70 per cent. or even over with super-atmospheric compression. This value has been adopted as an assured maximum in practice. On considera-tion of the extremely probable increased efficiency under sub-atmospheric conditions, a maximum possible value for the efficiency (under those conditions) will be taken at 75 per cent.

For a survey of the efficiencies of the various types of gas turbines, and their comparison with those of other prime movers, we have a double basis of calculation: one upon the maximum values of the limits in common present-day practice, and the other upon the very probable values which, with the conditions under which gas turbines would work, might not unlikely be achieved.

	T_2, non-water-cooled. °C	T_2, water-cooled. °C	u ft./sec.	e_t turbine.	e_p pump.
				Per cent.	Per cent.
Limiting values in common practice.	400	500	800	70	70
Limiting values under special conditions.	500	700	900	75	75

On this dual basis a survey of the economic position of the gas turbine is carried out in the next chapter.

CHAPTER IX. SUMMARY OF EFFICIENCIES AND COMPARISON OF TYPES.

ONLY three types of gas turbines have given, in practice, any promise of approaching the efficiencies achieved in other prime movers. These three are the single-fluid, constant-pressure, gas turbine; the single-fluid, constant-volume gas turbine; and the mixed-fluid, constant-pressure steam-and-gas turbine.

Of these three turbines, the first two have been built as actual working machines and experimented with (Armengaud and Lemale and Holzwarth); the third, has not, as far as the author is aware, been put to the test of actual experiment—at least, not with a machine of any size (with a *caveat* as to the claims of the Armengaud-Lemale turbine being considered in this category). It is now possible to make a general survey of these three types of gas turbines in respect to the efficiencies which, in practice, they might be expected to give.

In the last chapter it was pointed out how the efficiency of the gas turbine depended absolutely upon the limits of temperature, velocity, etc., within which it had to operate, and a dual list of these limits was laid down: the one such as could be entertained with some certainty from present-day practice in allied cases; the other, as might be expected (by one of a sanguine temperament) to be permissible under such special conditions as, in the case of the gas turbine, might possibly be achieved. These three types of gas turbines are to be considered under the following conditions:—

(A) By common practice:—

(1) Non-water-cooled—

$T_2 = 400° C.: u = 800$ ft./sec.

$e_t = e_p = 70$ per cent.

(2) Water-cooled—

$T_2 = 500° C.: u = 800$ ft./sec.

$e_t = e_p = 70$ per cent.

(B) By special conditions :—

(1) Non-water-cooled—

$T_2 = 500°$ C. : $u = 900$ ft./sec.

$e_t = e_p = 75$ per cent.

(2) Water-cooled—

$T_2 = 700°$ C. : $u = 900$ ft./sec.

$e_t = e_p = 75$ per cent.

This will give us the " overall " efficiency of the various types with the exception of such heat losses as may take place from the combustion chamber, no factors for which have been included in the above limits.

The omission of the factors for combustion chamber heat losses have not been included in the above limits for the reason that they are different for the different types of turbines. In the case of the constant-volume gas turbine, they would appear to vary from 40 per cent. (Clerk) to 15 per cent. (Holzwarth). They are here taken at 30 per cent. for common practice and 15 per cent. for special conditions, giving " furnace efficiencies " of 70 per cent. and 85 per cent. respectively. Similar figures have been chosen for the constant-pressure gas turbine, though it is reasonable to suppose that, in this case the losses would tend to be higher. First, the combustion chamber is continually subjected to the maximum temperature of the cycle, instead, as in the case of the explosion turbine, being only intermittently exposed to it; secondly, because an " air-to-air " regenerator would tend to serve as a centre for the distribution of radiant heat. In the explosion gas turbine, where the regenerator takes the form of a steam boiler, this source of heat loss is eliminated.

The type of explosion turbine chosen here for analysis is that of the Holzwarth type, the available waste heat from the turbine exhaust being used to drive the rotary pump for the initial compression, by the raising of steam and its subsequent expansion in a steam turbine. The range of expansion for this steam turbine is assumed to be 150 lbs. abs. to 1 lb. abs., giving a thermal efficiency of 28·3 per cent. (*vide* pp. 89 and 115). In reference to this, other factors influence the efficiency of the explosion turbine, the limit of peripheral velocity of 900 ft./sec., (velocity of fluid equals 4,500 ft./sec.) not being reached until the upper limit of temperature, T_1, which is fixed by r, the compression ratio of the pump, and T_2, the temperature after expansion, has been achieved. These limits are dealt with latter.

In the case of the steam-and-air turbine, the furnace, or combustion chamber, as well as the turbine itself, is completely water-jacketed. The losses by radiation and conduction from the furnace and regenerator are in the same category as those from a steam boiler.

In the following comparison of efficiencies all such losses have been taken at 10 per cent. for the case of common practice, and 5 per cent. for that of special conditions. In the case of the mixed-fluid turbine that type only has been considered as is likely to give the best results in practice, namely, the sub-atmospheric type. The derived type, with partial steam expansion on a separate wheel, is also considered.

Let it be assumed, to begin with, that there are no furnace losses.

CASE I.—THE SINGLE-FLUID, CONSTANT-PRESSURE GAS TURBINE.

A 1. By common practice, not water-cooled—

$T_2 = 400°$ C. : $u = 800$ ft./sec. : $e_t = e_p = 70$ per cent.
$t_2 = 215°$ C.

Then the velocity of the working fluid—

$$\begin{aligned} v &= n \,.\, u \\ &= 5 \,.\, 800 = 4{,}000 \text{ ft./sec.} \end{aligned}$$

and the work done—

$$W = \frac{4{,}000^2}{2g \,.\, J} = 177 \text{ T.U.}$$

Now

$$\begin{aligned} W &= C_p\,(\theta_1 - \theta_2) \\ &= C_p\,\theta_2\,(R^n - 1), \text{ where } n = (\gamma - 1)/\gamma. \end{aligned}$$

With $\theta_2 = 673°$ A,

$$R = 13{\cdot}5.$$

It is assumed that, under conditions of common practice, the exit temperature for the regenerator would not be likely to fall below 215° C. We have, therefore, waste heat from regenerator—

$$\begin{aligned} h &= C_p\,(T_2 - t_2) \\ &= C_p\,(215 - 15) \\ &= 50 \text{ T.U.} \end{aligned}$$

The negative work done by the pump—

$$\begin{aligned} w &= 18{\cdot}7 \log_e 13{\cdot}5, \\ &= 48{\cdot}6 \text{ T.U.} \end{aligned}$$

Therefore, thermal efficiency—

$$\eta = \frac{W - w}{W + h} = 56{\cdot}6 \text{ per cent.}$$

and overall efficiency—

$$\begin{aligned} E &= \delta\, \frac{{\cdot}7\, W - w/{\cdot}7}{W + h} \\ &= 24 \text{ per cent.} \end{aligned}$$

For the limits under condition A 2 (by common practice, but with water-cooling) $T_2 = 500°$ C., other limits as above, we have, from the same analysis—

$$\begin{aligned} \eta &= 59 \text{ per cent.} \\ E &= 27 \text{ per cent.} \end{aligned}$$

Similarly, for conditions of limits B 1, namely, $T_2 = 500°$ C.; $u = 900$ ft./sec. : $e_t = e_p = 75$ per cent., and with exit temperature from regenerator taken at 175° C., instead of 215° C.—

$$\begin{aligned} \eta &= 65 \text{ per cent.} \\ E &= 37{\cdot}3 \text{ per cent.} \end{aligned}$$

and for B 2, with limits as above, with the exception of $T_2 = 700°$ C.—

$$\begin{aligned} \eta &= 68 \text{ per cent.} \\ E &= 41{\cdot}2 \text{ per cent.} \end{aligned}$$

CASE II.—THE SINGLE-FLUID, CONSTANT-VOLUME GAS TURBINE.

The explosion turbine of the Holzwarth type is here chosen for analysis as representing the most practically successful type of constant-volume gas turbine. In fixing the limits within which this machine is to work, it is necessary to discard the peripheral velocity of the turbine wheel as a limiting factor in fixing the efficiency of the turbine. The reason of this is that, the value of θ_2 being fixed, a limiting value for θ_1 is reached before the work done upon this expansion from θ_1 to θ_2 has reached the limiting value, $W = (nu)^2/2gJ$, with the value of "u" from 800 to 900 ft./sec. The dependence of θ_1 on the new factor is shown as follows :—

Let R represent the expansion ratio in the turbine, and r that in the pump. Let the initial temperature be θ_1. Assuming the working fluid to obey Charles' law, we have—

$$R = r \,.\, \theta_1/\theta_0.$$

Now $\theta_1 = \theta_2 \,.\, R^{\frac{\gamma-1}{\gamma}}$: therefore—

$$\theta_1 = \theta_2 \,(r \,.\, \theta_1/\theta_0)^{\frac{\gamma-1}{\gamma}},$$

where θ is the atmosphere temperature.

Or—

$$\theta_1 = \theta_2^{\gamma} \,(r/\theta_0)^{(\gamma-1)}$$

It is thus seen that θ_2 being fixed, θ_1 depends upon the ratio of compression in the pump. Now, in the particular type of explosion chosen for consideration, it is assumed that all the available waste heat from the turbine exhaust (*i.e.*, from θ_2 to θ_0) is utilised in raising steam to drive the pump. It is also assumed that the thermal efficiency of the steam cycle adopted is 28·3 per cent. (*i.e.*, 150 – 1 lb., *vide supra*). Let the thermodynamic efficiency of the steam turbine driving the pump be denoted by e'', the efficiency of the regenerator raising the steam by e' ; the thermodynamic efficiency of the pump, as before, e_p, and that of the gas turbine, e_t.

The heat abstracted from the regenerator is done under conditions of constant pressure. The heat available is, therefore—

$$h = C_p \,(\theta_2 - \theta_0).$$

Therefore we have the negative work done by the pump in compression is—

$$w = h \,(e'' \,.\, e', \, e_p).$$

We have, also—

$$w = k \,.\, \log_e r$$

where k is a constant.

Therefore

$$\log_{\epsilon} r = \frac{h}{k} \cdot (e'' \cdot e' \cdot e_p).$$

This is a limiting value for the compression ratio in the pump dependent upon the values chosen for θ_2, e'', e', and e_p.

The value of the efficiency of the regenerator, e', has been taken as constant under all conditions at 80 per cent. This is in accord with the best steam boiler practice. For condition A of common practice the thermodynamic efficiency of the steam turbine, e'', has been taken at 70 per cent.; for condition B of special conditions at 75 per cent. The remainder of the limiting conditions are taken the same as for the constant-pressure, single-fluid gas turbine, for states of condition enumerated; the limits, u and t being omitted.

A 1. By common practice, non-water-cooled—

$$T_2 = 400° \text{ C.} : e_t = e_p = e'' = 70 \text{ per cent.} : e' = 80 \text{ per cent.}$$

With the explosion turbine, however, we have an added loss in the thermodynamic efficiency of the turbine owing to the variable velocity of the effluent gases on the turbine wheel. This loss, under conditions that "$u = v/3$" amounted to 18·6 per cent. of the total kinetic energy available (*vide*, Chapter V). Let it be assumed, in round figures, that this loss is 18 per cent. The value of e_t then becomes, on the same basis of comparison as in the case of the constant-pressure, single-fluid turbine (70 — 18) or 52 per cent. Therefore, emended value for e_t, is 52 per cent.

Now we have—

$$\theta_1 = \theta_2'^{\gamma} (r/\theta_0)^{1-\gamma}$$
$$= 1{,}340° \text{ A}$$

a value being taken here for θ_2' of mean between 673 and 773° A—for A 1—by common practice.

The work done—

$$W = \text{heat put in} - \text{heat rejected.}$$
$$= C_v (\theta_1 - \theta_0) - C_p (\theta_2 - \theta_0).$$

Substituting for C_p, $\gamma . C_v$, we have—

$$W = C_v (\theta_1 - \gamma\theta_2 + (\gamma - 1)\, \theta_0).$$

This gives value for work done ($\theta_2 = 673$)—

$$W = 68{\cdot}4 \text{ T.U.}$$

The heat put into the combustion chamber—

$$H = C_v (\theta_1 - \theta_0)$$
$$= 164 \text{ T.U.}$$

Therefore, thermal efficiency—

$$\eta = W/H.$$
$$= 68{\cdot}4/164 = 41{\cdot}7 \text{ per cent.}$$

Taking e_t at 52 per cent., we have overall efficiency—

$$E = 41{\cdot}7 \text{ per cent.} \times 52 \text{ per cent.}$$
$$= 21{\cdot}7 \text{ per cent.}$$

Similarly, for condition A 2, $T_2 = 500°$ C., other limits as before (θ_2' still at mean value between 673 and 773) we have—

$$\eta = 42{\cdot}6 \text{ per cent.}$$
$$E = 22{\cdot}2 \text{ per cent.}$$

The same methods of calculation are employed for condition B, when $T_2 = 500°$ C. (1) or 700° C. (2), and $e_t = e_p = e'' =$ 75 per cent. : $e' = 80$ per cent., θ_2' being here taken at the mean between 773 and 973° A. We have then for condition B 1—

$$\eta = 51 \text{ per cent.}$$
$$E = 29 \text{ per cent.}$$

And, for condition B 2—

$$\eta = 52{\cdot}6 \text{ per cent.}$$
$$E = 30 \text{ per cent.}$$

CASE III.—THE MIXED-FLUID CONSTANT-PRESSURE GAS TURBINE.

(A) Sub-atmospheric Steam-and-Gas Turbine.

With this form of gas turbine, as with the single-fluid, constant-pressure type, the value of R, θ_2 having been fixed, is determined by the value of u, the peripheral speed of the turbine wheel. In this instance, however, as has been pointed out in the preceding chapter, it is not possible to determine R directly from the equation for work done, as the equation in question involves dependent variables, such as y, the weight of steam added, and the factors in the Rankine formula for the steam expansion. Arbitrary values for R have to be fixed from which the value of u for that particular case can be obtained. If this value is above the value required for the limiting conditions considered, another value for R must be chosen which will give a value for u below the limiting one.

We thus have two values for the overall efficiency, one, E_1, corresponding to peripheral speed limit $(u + x)$, and another, E_2, corresponding to a peripheral speed limit $(u - y)$, where x and y are small compared to u. Assuming that the change in E over the range $(u - y)$ to $(u + x)$ to be represented by a linear equation, it is easy to obtain a value for E when peripheral velocity is u, by means of a straight line graph. This gives a close approximation to the absolute value of E corresponding to u.

Limiting conditions : A 1. By common practice, non-water-cooled—

$$T_2 = 400^\circ \text{ C.} : e_t = e_p = 70 \text{ per cent.} : u = 800 \text{ ft./sec.}$$

Let the value of R be taken at 10 ; the turbine being assumed to operate between the pressures of 15 lbs. and 1·5 lbs. abs.

Now we have that the initial temperature of the steam, θ_1', is given by the equation—

$$\theta_1' = \theta_2 . \text{R where } \gamma = 1{\cdot}3.$$

Therefore—

$$\theta_1' = 1{,}150^\circ \text{ A.}$$

Similarly, for the products of combustion—

$$\theta_1 = \theta_2 . \text{R where } \gamma = 1{\cdot}38.$$

Therefore

$$\theta_1 = 1{,}270^\circ \text{ A.}$$

The work done by 1 lb. of the products of combustion—

$$\begin{aligned} W_a &= C_p\ (\theta_1 - \theta_2). \\ &= 150 \text{ T.U.} \end{aligned}$$

The work done by 1 lb. of the steam, by the Rankine formula—

$$W_s = (\tau_1 - \tau_2)\left(1 + \frac{L_1}{T_1}\right) + C_{ps}\ (\tau_s - \tau_1) - \tau_2\left(\log_e \frac{\tau_1}{\tau_2} + C_{ps} \log_e \frac{\tau_s}{\tau_1}\right).$$

The maximum quantity of steam added to the products of combustion is found by the usual formula (*vide* Chapter III.)—

$$y . m = H - C_{pa}\ (\theta_1 - \theta_0) + (t_1 - t_2)\ (C_{pa} + y . C_{ps}).$$

This gives a value for y of ·444 lb. We have, then, the work done by the steam added to the products of combustion, equals

$$W_s \cdot y = 128 \text{ T.U.}$$
$$W_a = 150 \quad ,,$$

Therefore $W_{total} = 278$,,

The negative work—

$w = 18{\cdot}7 \log_e 10$
$= 43$ T.U.

The heat put in—

H = 550 T.U.

(the heat value for 1 lb. of products of combustion). Therefore—

$\eta = (W - w)/H = 42{\cdot}8$ per cent.

And overall efficiency—

$$E = \frac{{\cdot}7\ W - w/{\cdot}7}{H}$$

$= 24{\cdot}4$ per cent.

Now the work done here per lb. of *working fluid* is—

278/1·444 T.U.

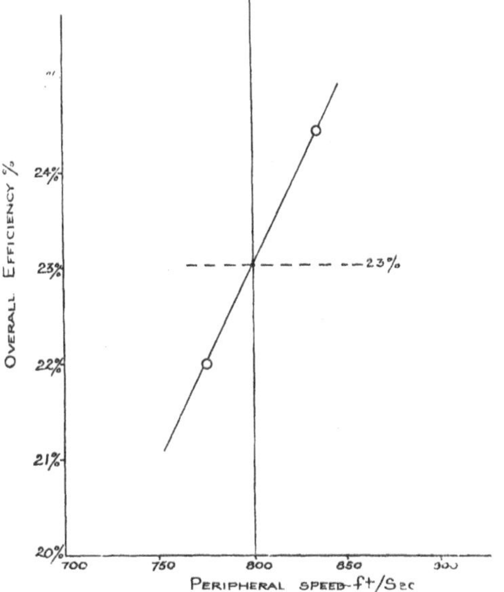

FIG. 84.—Speed-Efficiency Graph for Mixed-Fluid Turbine.

or 193 T.U. This value gives a value of 4,180 ft./sec. for the velocity of the gases on the turbine wheel, and, if the factor of reduction be taken at 5, a value for the peripheral velocity, u, of 836 feet per second.

It is thus seen that the value chosen of R was too high for the peripheral speed of 800 ft./sec. Let the same case be worked out for a value of R of 7·5, that is, $p_2 = 2$ lbs. This gives values—

$$\eta = 38{\cdot}7 \text{ per cent.}$$
$$E = 22{\cdot}0 \text{ per cent.}$$

And a value of $v = 3{,}880$ ft./sec; therefore, $u = 776$ ft./sec.

In Fig. 84 a straight line graph is plotted for E and u from these two cases, giving a value, when u equals 800 feet per second, of—

$$E = 23 \text{ per cent.}$$

Similarly, for conditions as above, but with T_2 equal to 500° C. instead of 400° C.—

$$E = 24{\cdot}5 \text{ per cent.}$$

For case B 1 (special conditions, non-water-cooled) $T_2 =$ 500° C. : $u = 900$ ft./sec., $e_t = e_p = 75$ per cent.—

$$\eta = 49 \text{ per cent.}$$
$$E = 32{\cdot}3 \text{ per cent.}$$

And for case B 2 (special conditions, water-cooled) $T_2 =$ 700° C. : $u = 900$ ft./sec., $e_t = e_p = 75$ per cent.—

$$E = 31{\cdot}5 \text{ per cent.}$$

(B) Derived Type : Mixed-fluid, Sub-atmospheric ; Separate Steam Expansion on High-pressure Wheel.

In this case T_2 is taken at 500° C. (water-cooled or otherwise) u at 900 ft./sec., and $e_t = e_p = 70$ per cent. (1) ; $e_t = e_p = 75$ per cent. (2).

The thermal efficiency for this cycle is—

$\eta = 56{\cdot}6$ per cent. (*vide* p. 93).

This gives values for overall efficiency—

(A) $E = 34$ per cent.
(B) $E = 38$ per cent.

We are now able to make a *précis* of the results so far obtained on a comparative basis as follows :—

(A) Single-fluid Turbines.

	Values of "E" (per cent.). (Combustion chamber heat losses neglected.)	
	Non-water-cooled.	Water-cooled.
I. Constant-pressure :		
By common practice	24·0	27·0
By special conditions	37·3	41·2
II. Constant-volume :		
By common practice	21·6	22·1
By special conditions	29·0	30·0

(B) Mixed-fluid Turbines.

	Non-water-cooled.	Water-cooled.
I. Sub-atmospheric, steam-and-gas turbine :		
By common practice . .	23·0 ..	24·5
By special conditions . .	32·3 ..	36·2
II. Mixed-fluid, sub-atmospheric ; steam wheel, super-atmospheric :		
By common practice . .	34·0	
By special conditions . .	38·0	

On this basis of comparison, then, the following is deduced : Under conditions of limitation, as at present exemplified in common practice, the derived type of constant-pressure, mixed-fluid turbine, with a separate high-pressure steam wheel, gives the highest promise of overall efficiency ; that even on the basis of " special conditions " this type still heads the list with an overall efficiency of 38 per cent. ; for it must be borne in mind that this value is calculated for those limitations which, in the constant-pressure, single-fluid gas turbine, give an overall efficiency of 37·3 per cent. It may be noticed that the superiority is more marked under conditions of common practice than under special conditions.

Next to this type comes the constant-pressure, single-fluid gas turbine, with efficiencies varying from 24 per cent. to 41·2 per cent. The mixed-fluid, steam-and-gas turbine comes next, giving values of E from 23 per cent. to 36·2 per cent. ; and, lastly, the explosion turbine, with efficiencies from 21·6 per cent. to 30 per cent.

One important factor in the fuel economy, however, has not yet been taken into account, namely, the heat losses from the combustion chamber. Let the combustion chamber efficiency be taken into account upon the same dual basis as that upon which the above values of overall efficiency have been based. It must be remembered that for different types of turbines the value of the heat loss varies (*vide supra*, p. 188). Let the combustion chamber efficiencies be taken as follows :—

(A) For single-fluid gas turbines (either constant-pressure or constant-volume) combustion chamber efficiency, e_x, equals—

(1) By common practice, 70 per cent.
(2) By special conditions, 85 per cent.

(B) For mixed-fluid turbines combustion chamber efficiency equals—

(1) By common practice, 90 per cent.
(2) By special conditions, 95 per cent.

We can then obtain the values of the total overall efficiency, *i.e.*, actual brake horse-power over total heat entering system per given unit of time, all losses between entrance of fuel to turbine and useful work on turbine shaft being taken into account.

(A) SINGLE-FLUID TURBINES.

	Values of "E_T" (per cent.). Heat loses from combustion chamber included.	
I. Constant-pressure :	Non-water-cooled.	Water-cooled.
By common practice . .	16·8 ..	18·9
By special conditions . .	31·7 ..	35·0
II. Constant-volume :		
By common practice . .	15·1 ..	15·5
By special conditions . .	24·6 ..	25·5

(B) MIXED-FLUID TURBINES.

I. Sub-atmospheric, without separate steam wheel :	Non-water-cooled.	Water-cooled.
By common practice . .	20·7 ..	22·1
By special conditions . .	30·7 ..	34·5
II. With partial super-atmospheric expansion on separate steam wheel :		
By common practice . .	30·6	
By special conditions . .	36·1	

This shows the comparative values of the various types of gas turbines treated of. The mixed-fluid turbine with high-pressure steam expansion on a separate wheel still gives the best final efficiency (30·6 to 36·1 per cent.); the sub-atmospheric mixed-fluid turbine comes next, easily outdistancing the single-fluid, constant-pressure turbine on the basis of common practice, while but barely reaching it on the grounds of "special conditions" (20·7 to 34·5 per cent.) (*vide* efficiency curves,

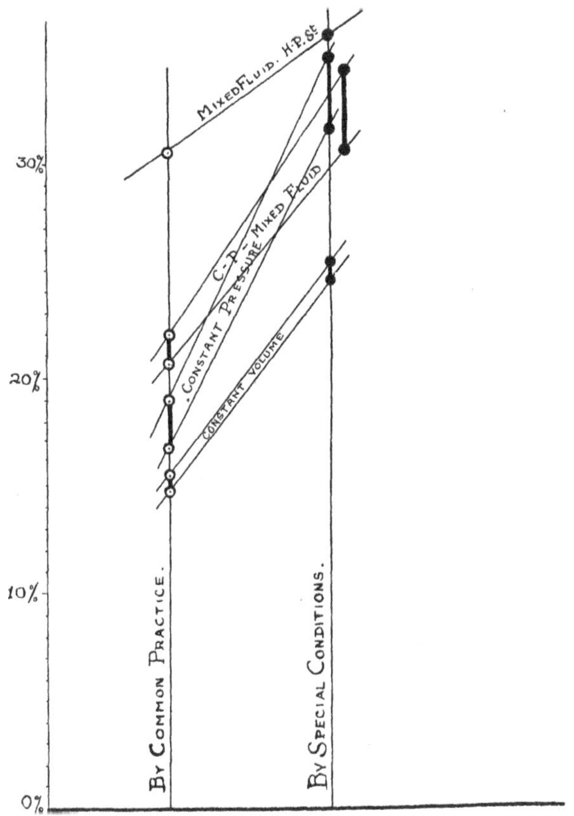

Fig. 85.—Graphic Representation of Overall Efficiencies of various Types of Gas Turbines.

Fig. 34). The single-fluid, constant-pressure turbine comes next (16·8 to 35 per cent.); and the single-fluid, constant-volume turbine comes last with final efficiencies of from 15·1 to 25·5 per cent.

The result of this analysis is shown graphically in Fig. 85. The values for the overall final efficiencies read on the vertical;

the values for E_T for common practice read along the left-hand line; those for special conditions along the right-hand line. Each type shows double transverse lines for the two cases of non-water-cooled blades and water-cooled blades, with the exception of the line for the mixed-fluid turbine with high-pressure steam wheel, for which only one value of T_2 was taken.

FUEL CONSUMPTION.

There are two cases to consider in this connection: one when the fuel is oil, the other when it is producer gas from coal.

(A) Fuel Consumption with Oil as Fuel.

Let the calorific value of the oil be 10,000 T.U. per lb. The values of oil consumption in terms of lbs. of oil per brake horse-power per hour are as follows :—

(A) Single-fluid Turbines.

		Lbs. of Oils.	
		By common practice.	By special conditions.
I. Constant-pressure :	1	0·842	0·446
	2	0·748	0·404
II. Constant-volume :	1	0·937	0·575
	2	0·914	0·555

(B) Mixed-fluid Turbines.

		By common practice.	By special conditions.
I. Sub-atmospheric, no high-pressure steam wheel :	1	0·683	0·461
	2	0·640	0·410
II. Sub-atmospheric, with high-pressure wheel :		0·462	0·392

(B) Fuel Consumption in Coal used per B.H.P., with Producer Gas as Combustible.

From the results of trials by Mr. H. A. Humphreys on the producer plant of Messrs. Brunner, Mond & Co., at Winnington, an efficiency was obtained of 84 per cent. (Robinson's "Gas Engine," p. 582). This may be taken in the category, B, of "special conditions." On the basis of common practice the producer efficiency is taken at 75 per cent. The gases, however, in the case instanced above left the producer at a temperature of 450° C. This represents an approximate heat loss of 4·5 per cent. In the case of the mixed-fluid, sub-atmospheric steam-and-gas turbine it is possible to conduct these gases straight to the furnace without any initial cooling, as here there is no initial compression to be done. How far such a scheme is practicable is very doubtful (*vide* p. 153); but, in dealing with probabilities, it is another point in favour of the mixed-fluid turbine, sub-atmospheric type. The increased value of the producer efficiency to 88·5 per cent. will not be considered in the following analysis of fuel consumption to be on the same basis of probability as the 84 per cent. "special condition" value, and this latter figure will be retained throughout. Figures for fuel consumption, however, at the 88·5 per cent. value, will be placed in brackets alongside the 84 per cent. values for the mixed-fluid types of turbine. The calorific value of the coal has been taken at 7,000 T.U. per lb.—for good Nottinghamshire slack. We have, therefore, the weight of coal consumed per brake horse-power per hour—

$$m = \frac{\cdot 202}{E_T \, . \, K},$$

where K is the efficiency of the producer.

The following table gives the fuel consumption in lbs. of coal per brake horse-power per hour for various types of turbines :—

(A) Single-fluid Turbines.

		Lbs. of Coal. By common practice.	Lbs. of Coal. By special conditions.
I. Constant-pressure :	1	1·600	0·761
	2	1·420	0·687
II. Constant-volume :	1	1·785	0·977
	2	1·740	0·943

(B) Mixed-fluid Turbines.

		By common practice.	By special conditions.
I. Sub-atmospheric, no high-pressure steam wheel :	1	1·305	0·783 (0·743)
	2	1·220	0·696 (0·661)
II. Sub-atmospheric, with high-pressure steam wheel :		0·877	0·665 (0·632)

ACTUAL COST BY FUEL CONSUMPTION, IN PENCE PER BRAKE HORSE-POWER PER HOUR.

The oil used for the Diesel engine may be taken at a cost of 50*s.* per ton delivered in England. The cost of the coal—Nottinghamshire slack, of 7,000 T.U. per lb., delivered within a reasonable distance of the coal fields, *i.e.*, in the north of England generally—may be taken at 10*s.* per ton. This affords the usual cost ratio of oil to coal fuel of 5 to 1. This gives the cost of oil per lb. at 0·268*d.*, and cost of coal per lb. at 0·0536*d.* The costs for oil and coal fuel, in pence per brake horse-power per hour, are given in the following columns. The symbols, A 1, A 2, B 1, B 2, refer to the conditions detailed on pp. 187, 188, *et seq.*

	Pence per Brake Horse-power per Hour. Oil Fuel.		Coal Fuel.
Single-fluid, constant-pressure :	*d.*		*d.*
A 1 . . .	0·226	..	0·086
2 . . .	0·220	..	0·076
B 1 . . .	0·120	..	0·041
2 . . .	0·108	..	0·037
Single-fluid, constant-volume :			
A 1 . . .	0·251	..	0·096
2 . . .	0·245	..	0·094
B 1 . . .	0·154	..	0·053
2 . . .	0·149	..	0·051
Mixed-fluid, sub-atmospheric :			
A 1 . . .	0·183	..	0·070
2 . . .	0·171	..	0·066
B 1 . . .	0·122	..	0·042 (0·040)
2 . . .	0·110	..	0·038 (0·036)
Mixed-fluid, with high-pressure steam wheel :			
A . . .	0·124	..	0·047
B . . .	0·105	..	0·036 (0·034)

We are now in a position to compare the gas turbine with other prime movers on an economic basis. The fuel consumption of other prime movers can be correlated with the tables for the gas turbine given above in the following way : " A " refers to average economies in common practice ; " B " to the best " test " results so far obtained. Thus we have—

Fuel Cost Diagram.

Comparison of Gas Turbine with other Prime Movers.

Upper Values – By Common Practice.
Lower Values – By Special Conditions.

PENCE PER B.H.P. PER HOUR

·240
·220
·200
·180
·160
·140
·120
·100
·080
·060
·040
·020
0

CONSTANT VOLUME OIL TURBINE.
CONSTANT PRESSURE OIL TURBINE.
STEAM AND OIL TURBINE.
S AND O.T. H P WHEEL.
DIESEL OIL ENGINE.
STEAM TURBINE.
STEAM ENGINE.
GAS ENGINE.
C-V: GAS TURBINE.
C P GAS T.
STEAM & GAS T.
S & G.H.P WHEEL

OIL
COAL

	E_T per cent.	Boil. or prod. eff. per cent.	Final eff. of plant. per cent.	Lbs. Fuel. 1. Coal.	Cost/B.H.P. Pence.
Steam turbine : A .	16	70	11·2	1·80	0·096
B .	21	80	16·8	1·20	0·064
Steam engine : A .	18	70	12·6	1·60	0·086
B .	23	80	18·4	1·10	0·059
Gas engine : A .	23	75	17·3	1·16	0·062
B .	28	84	23·5	0·86	0·046
				2. Oil.	
Oil engine (Diesel) :					
A .	30	—	—	0·472	0·127
B .	35	—	—	0·405	0·107

A general comparison of the cost in fuel consumption for the gas (oil) turbine, and other prime movers, is shown graphically in the cost chart (Fig. 86). The cost in pence per brake horsepower is read vertically, the figures being taken from the preceding columns for the various types of gas turbines and for other prime movers.

It is necessary to point out one fact of some significance in respect to the basis of cost comparison. In the case of the steam-and-gas (or oil) turbine there is a *greater degree of probability* in this type of attaining to the "special conditions" state of limitations than with other types of gas turbine (that is, T_2 may more easily be raised, owing to ease in water-cooling; e_p and e_t may more easily be increased, owing to decrease in friction in sub-atmospheric conditions (*vide* Chapters III., IV. and VII.)). The values of cost of fuel consumption do not, therefore, exactly correspond; the values for the mixed-fluid turbines are, in effect, slightly too high. It may be noted that the dotted line in the right-hand bottom corner of the chart indicates the locus of the costs of fuel consumption for the gas and steam turbine using the gas from the producer without initial cooling. This concludes the survey of the gas turbine from the standpoint of fuel economy.

CHAPTER X. THE HISTORY OF THE GAS TURBINE.

"Etenim per conjunctionem æris, ignis, aquæ & terræ: & tribus elementus, vel etiam quattuor inter se congredientibus variæ dispositiones efficiunter; aliæ quidem necessarios vitæ nostræ usus præbentes; aliæ vero terribilem quandam admirationem moventes"—*Heronis Alexandrini Spiritalia* ("Veterum Mathematicorum," Thevenot, Parisiis. MDCXCIII.).

About 130 years before the birth of Christ there lived at Alexandria, in the reign of Ptolemy Philadelphus, a certain Hero, who "was eminently distinguished in that age and region of refinement, not only for the extent of his attainments in the learning of his time, but also for the number and ingenuity of his mechanical inventions." It is well known this Alexandrian mathematician was responsible for the first steam engine. The form that it took was that of the "æliopile"—or steam reaction wheel—and "held the market" among heat engines for some 1,700 years. It may be, however, not so widely known that the first "per conjunctionem æris, ignis, aquæ et terræ . . . variæ dispositiones *efficire*" was also the first inventor of the gas turbine. That is to say, the first to apply the expansive properties of air for the production of kinetic energy, and the subsequent absorption of that energy upon a rotating wheel.

The machine which Hero devised was for the purpose of causing certain figures, placed below an altar, to be set in motion as soon as the altar fire was lit. His description, from the *Spiritalia*, is as follows:—

"If, in another kind of altar, the fire is lighted, living figures may be seen to lead the chorus; for if made of glass or horn, these altars are transparent. A tube is led through the hearth above down to the base of the altar, which rotates about a pivot and the mouth of the tube, the upper part of which is fixed in the hearth. But the large tube has, also, other small, bent tubes joined on to it, which are at the same time perforated, fixed at two diameters, and themselves bent in turn. Moreover, a wheel is fixed to these upon which are placed the live figures that lead the chorus. Thus, as soon as the altar fire is lit, the air becomes heated, and passes through the mouth of the tube into the tube itself; from this it is expelled through the small bent tubes and pressing up against the hollowness of the altar, turns the tube and the live figures which lead the chorus."

The general construction of this machine is shown in Fig. 87, taken from Thevenot's "Veterum Mathematicorum" edition of the *Spiritulia*. It will be seen that the apparatus works, not

FIG. 87.—Hero's Gas Turbine (from Thevenot's "Veterum Mathematicorum").

by the expulsion of air from the extremities of the bent tubes as in the Barker's mill, but by the indrawing of the air through these tubes brought about by the rarefaction of the air in that part of the large tube which is heated by the fire. It would seem that either Hero mistook the theory upon which his machine worked or, what is more likely, that the seventeenth century Latin translators misconstrued the original Greek codices. The diction used is spoken of in the preface as "Pneumatici discriptiones quæ omnibus Interpretibus adea difficilis visæ sunt."

The machine, in fact, is an impulse, and not a reaction

turbine; it is precisely similar in operation to that which became known later as the "smoke jack."

The next step in the evolution of the heat engine was the invention, by Giovanni Branca, of the steam impulse wheel, at the beginning of the seventeenth century, (essentially) the forerunner of the De Laval turbine. It seems, however, that the gas turbine even preceded this, for the same mechanism was described, before this period, "by Cardan as moved by the 'vapour from fire.'"[1] The next notice of the theory of applying the expansive force of air directly to produce rotary motion was Amonton's "fire wheel," in 1699. This ingenious machine, however, can hardly be regarded as a "gas turbine" proper, it being in reality a "gravity wheel" in which water was displaced along the wheel by the expansive force of heated air.

The first notice of a gas turbine bearing the essential features of such machines to-day came nearly 100 years later, when John Barber, "of Attleborough, in the parish of Nuneaton, in the county of Warick, Gentleman," obtained a patent for the first gas engine, or, to be more accurate, for the first gas turbine. It is a matter of some significance to note that the first steam engine known, namely, Hero's steam reaction wheel, was, in effect, a steam turbine; and the first gas engine, namely, that of John Barber, was a gas turbine. Between the date of Branca's impulse wheel at the beginning of the seventeenth century and the filing of Parson's first patent in 1884, over 260 years had elapsed. During that period the evolution of the steam engine proceeded along other lines. But the evolved product for the use of steam as a motive fluid is precisely the same in its initial shape, viz., the steam turbine.

Between the first gas engine, or *rather gas turbine*, of John Barber, in 1791, and the explosion gas turbine of Holzwarth, in 1911, 120 years have elapsed. During that time the gas engine has been evolved along other lines. To-day it seems as if a return were about to be made to the initial type; and the final evolved product of the use of gas as a motive fluid may revert to the early type, as set forth by Barber, by Dumbell, by Bresson and others.

It is interesting to note that Barber's gas turbine contained

[1] "History of the Steam Engine," Stuart, 1824, p. 6.

all the essential features of the gas turbine of to-day. There was the compressor for compressing the gas, and a similar one for compressing the air ; an exploder, in which the gas and air was introduced and exploded. Moreover, water was introduced into the combustion chamber, " to prevent the inward pipes and the mouth of the exploder for[1] melting by the velocity and intenseness of the issuing flame." This specification of Barber's is so significant in the light of modern gas turbine practice that it will well bear quotation in full, as follows :—

" NOW KNOW YE, that I, the said John Barber, in compliance with the said proviso, do hereby describe the nature of the aforesaid Invention, and declare that the same is to be performed in manner following, that is to say :—

" 'It consists of a metallic vessel called a retort, so contrived that (when heated by a circumambient fire), coal, wood, oil, or any other combustible matter may be put therein, and the smoke or vapour therein collected may be brought out by a small pipe, and conveyed in a regular stream into another metallic vessel called an exploder, by means of an air pump and a compresser regulating bellows, which pipe opposing its orifice to another similar pipe which enters the exploder (on the opposite side from the pipe which brings in the inflammable vapour from the retort) and injects by similar means a proper quantity of atmospheric or common air, causing an admixture of the two airs, which, so mixed, will take fire on application of a match or candle to the mouth of the exploder, and rush out with great rapidity in one continued stream of fire, so long as the exploder is supplied with proper quantities of the respective airs. The fluid stream is also considerably augmented both in quantity and velocity by water injected into the exploder by means of another small pipe entering therein, which water is also intended to prevent the inward pipes and the mouth of the exploder for[1] melting by the velocity and intenseness of the issuing flame. This water, as well as the airs, are forced into the exploder by means of a pump, which in lieu of common crank carries upon the axis of one of its wheels two esses or double portions of circles, whereby a more regular motion is procured than can be done by any crank work. This engine is wrought by the steam issuing from the mouth of the exploder, and may be applied to grinding, rolling, forging, spinning, and every other mechanical operation ; and the fluid stream may be injected into furnaces for smelting metallic ores, or passed out at the stern of any ship, boat, barge, or other vessel, so as by an opposing and impelling power directed against the water carrying such vessel, the vessel with its contents may be driven in any direction whatsoever.'

" In witness whereof, I, the said John Barber, have hereunto set

[1] *I.e.*, "from": "for" is the correct reading of the 1791 text.

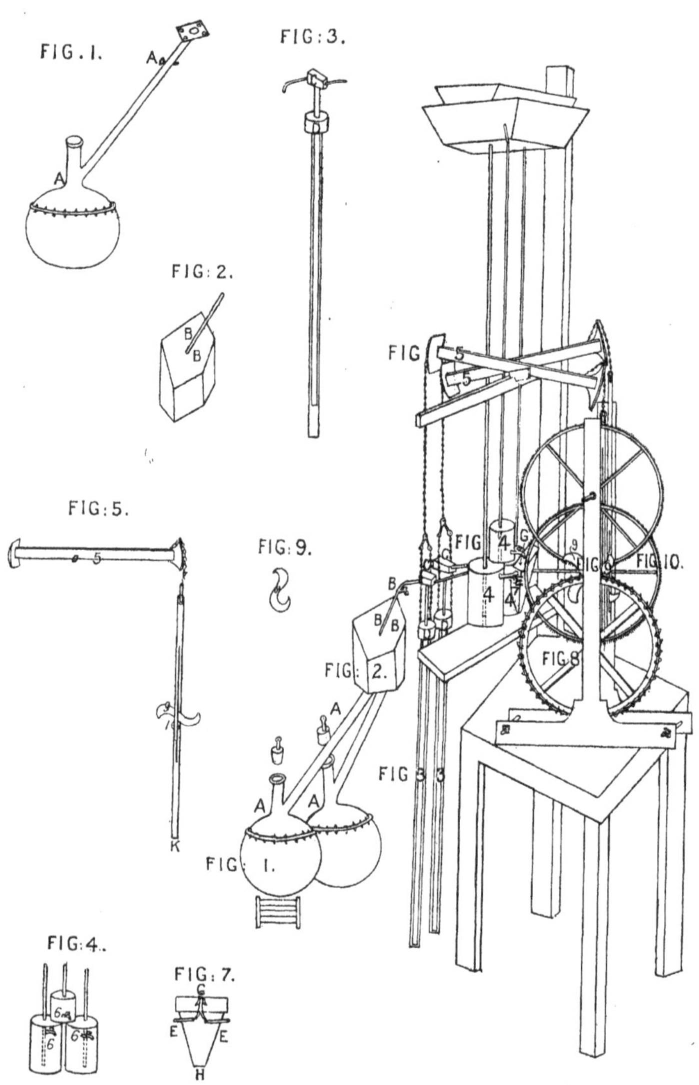

FIG. 88.—Barber's Gas Turbine. (From Barber's Specification, No. 1,833, A.D. 1791.)

A SPECIFICATION OF AN ENGINE FOR USING INFLAMMABLE AIR FOR THE PURPOSE OF PROCURING MOTION AND FACILITATING METALLURGICAL OPERATIONS, AND WHICH MAY BE APPLIED TO THE GRINDING OF CORN, FLINT, MANGANESE OR OTHER MATTER, ALSO TO ROLLING, SLITTING, FORGING AND BATTERING IRON AND OTHER METALS, TURNING OF MILLS FOR SPINNING AND ENGINES FOR TURNING UP COALS, MINERALS FROM MINES OF ALL SORTS, STAMPING OF ORES, RAISING WATER, AND ANY OTHER MOTION THAT MAY BE REQUIRED.

FIGURE 1. Is a Retort with its Neck (A) and a Stop Cock (A). I make use of two of these Retorts in order to keep the Engine working with one of them whilst the other is cleansed of the Coaks and Ashes of the Materials used for procuring Inflammable Air.

FIGURE 2. Is a Receiver wherein the Air from either of the Retorts is collected and cooled by means of a Circurmumbrient Cistern of water B B, and carried by a small pipe into the Pump for supplying the Engine with Inflammable Air.

FIGURES 3, 3a. Are the Air pumps which by means of an Admission Valve through which the respective Airs are to be drawn into the Cistern of the Inward Pipe C C (left open at the bottom), and another Valve where the Airs are to pass out of the Cistern into the Compressors, does upon the alternate raising and falling of the outward pipes D D (filled with water) convey the Airs into their respective Compressors by being successively raised and fallen by means of two Beams 5, 5.

FIGURES 4, 4, 4. Are Compressors which being filled in their lower parts with water and the respective Airs forced into them by means of their adjoining Pumps, do, according to the altitude of the Pipes rising up to their Cistern, compress the Air and force it into the Exploders with a power proportioned to such altitude. In some cases I chose to use forcing Pumps and Bellows for performing this Operation.

FIGURES 5, 5. Two Beams wrought by two Esses or portions of Circles fixed on one of the Axis's of the Engine Wheel.

FIGURES 6, 6, 6.[1] Three Stop Cocks to regulate the Admission of Inflammable Air, Common Air and Water, into the Exploder.

FIGURE 7. The Exploder with a Pipe (E) which introduces Inflammable Air, a Pipe (F) which introduces Common Air and a Pipe (G) which introduces Water into this Vessel, at whose Mouth or outlet (H) (upon the approach of Flame) the combined Streams of Airs and Water do issue out with amazing force and velocity against

FIGURE 8. The Fly Wheel on whose Axis a Pinion being placed turns a toothed Wheel of proper dimensions and on the same Axis are affixed

FIGURES 9, 9. Two Metallic Esses or Portions of Circles set at right Angles from each other so as by rotation of the Axis whereon they be fastened they bear upon the Wheel (I I)[1] in the Plug frame (K K)[1] suspended to the Beams 5 5, and thereby work those Beams two strokes each in every rotation.

FIGURE 10. Is the second Wheel in the Engine wrought by the Pinion upon the Fly Wheel, and upon whose Axis is fixed another Pinion which working upon another toothed Wheel of any proper diameter, may upon the Axis thereof carry either Cogs or Esses for lifting Hammers, Barrels for Ropes to wrap upon Cog Wheels for communicating Motion to Grinding or Spinning Machinery, or rolling, slitting and pressing Metals or other substances, and also for any kind of Motion whatsoever.

Any sort of Engines and Machines worked upon these principles may be erected at moderate expense, wrought with very little fuel, and managed with little Trouble in comparison with any other Machines hitherto invented.

[1] These references do not occur in the illustration; the above notes are a word for word reprint from the original specification. It has been thought best to reproduce the text and illustrations of the 1791 patent, without alteration.

my hand and seal, this Twenty-eighth day of November, in the year of our Lord One thousand seven hundred and ninety-one, and in the Thirty-second year of the reign of our Sovereign Lord George the Third, by the Grace of God of Great Britain, France, and Ireland King, Defender of the Faith, and so forth.

"JON. (L.S.) BARBER."

In Fig. 88 is shown a facsimile of the drawing accompanying Barber's specification; the notes facing it are those attached to Barber's drawing. The last note, it may be observed, is of almost too Elysian a nature even for the most enthusiastic of inventors to-day.

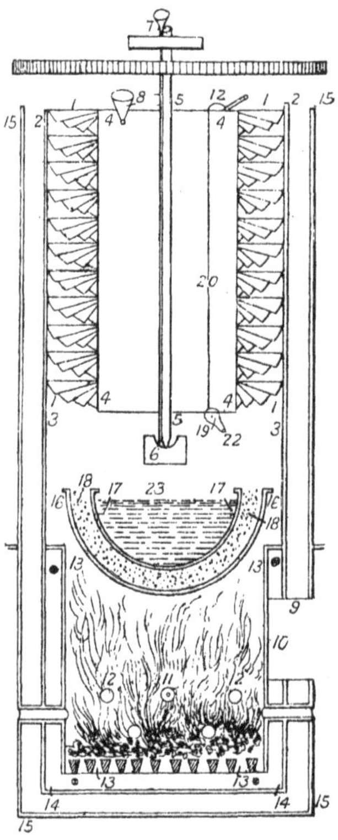

FIG. 89.—Dumbell's Gas Turbine.

After Barber's patent in 1791 there was no further advance made in respect to the gas turbine until the year 1808 when John Dumbell, "of Mersey Mills, in the Parish of Warrington, and the County Palatine of Lancaster, Miller," obtained a patent entitled, "Obtaining and Applying Motive Power, the Construction of Carriages, etc." The specification possesses eighteen separate figures, together with fifteen pages of text, and the title certainly has sufficient latitude for such. But the only figure with which we are here concerned is that marked "Fig. A" in the specification and reproduced here (Fig. 89).

The machine is described in the inventor's own words as follows :—

"1st, Figure A is a section of one kind of apparatus. No. 1 is a vessel open at top, and cylindrical from No. 2 to No. 3 (the shape of its other parts, ad libitum), in which is the interior cylinder, No. 4, closed, or partially closed, both at top and bottom; the outer side or circumference of this interior cylinder has any required number of flyers (made of metal when so required, and either spirally or horizontally) affixed to it according to its length. In the Figure A eleven of these flyers are attached, which are made to resemble the sails of a windmill or the flyers of a smoke jack; and where more than one is used they are placed stratum super stratum (in some cases cones or vessels of other shapes may be used); this interior cylinder No. 4 turns round by means of the spindle No. 5, and on the step No. 6, which step is supported from the sides of the cylinder or otherwise, and is from time to time kept cool by means of a supply of grease, &c., introduced through a tube resembling a gun barrel No. 7. The interior cylinder No. 4 may be used as an empty cylinder, or may be filled full, or partially full, of water or other fluid, body, or thing, through an orifice at the top, as at the funnel No. 8, in which latter cases the weight of such fluid or thing may act as a fly or regulator; and such water might counteract colliquefaction, or any injurious effect of the fire beneath. No. 9 is the manhole, made to open as a door, and fitted up with hinges and latches; or it may be fastened down as is usual in a steam engine boiler, and used for similar purposes as at a steam engine, as well as for occasionally charging, discharging, supplying, and cleaning the receptacle. No. 10 is the fire-feeding hole on the manhole lid, with hinges and latches, or otherwise, as above mentioned. No. 11 is the bellows hole, at which place a tewel for a bellows pipe, air pump, or conductor may be affixed, and through which may be sent in, by compression or otherwise, suitable air for ignition, or inflammable gas, or such other ingredient or body as may be most fit to fan a flame or to encourage flammability; the use of bellows for this purpose may be perhaps best understood by those who have witnessed the effect of a watchmaker's blowpipe; and until the fire is sufficiently ignited to work the bellows by the action and power of the machinery, one or more of them (which may be of a cylindrical form or otherwise) may be worked by manual labour during such insufficiency.

. . . .

"No. 13 is the fireplace. No. 14 is the bottom part of the cylinder surrounding the fireplace and the other apparatus; between the grate and the cylinder is a receptacle for air, cinders, &c. marked ⊗, ⊗, ⊗, ⊗. No. 15 is an outside cylinder or water pipe, thro' which pipes are to pass to admit the air at No. 12, in order that the space between Nos. 14 and 15 may be filled, when required, with water or some other intervening body. No. 16 is a pan or boiler, receiving its heat from the fire in the grate No. 13. No. 17 is an interior pan or boiler (No. 18 forming a kind of balneum for No. 17); the space No. 18 between the two pans being filled with sand, water, or other matter as a retainer and conveyor of heat, and also to counteract the colliquefaction of the said pans."

Dumbell asserts later in his specification that he "would apply whatever produces steam, smoke, rarified air, or spirit, or which produces an inflammable vapour, either in a combined or single state, or which may be drawn by distillation, or which possesses an explosive or combustible power or effect."

This practically means that Dumbell's machine—which, it appears, he proposed to use for locomotion—was neither more nor less than the prototype of the "explosion" gas turbine; while Barber's machine, may more properly be considered as the forerunner of the constant-pressure gas turbines.

It is interesting to note that Dumbell figures in his specification a "compound" wheel, whereas Barber uses a single-impulse wheel; the absence of any guide blading, however, in the former case shows that no grasp of the principle of the compound turbine had yet been achieved.

On April 20th in the first year of the reign of Queen Victoria, a certain M. Bresson patented in Paris a machine for "heating and compressing air." The air was compressed and delivered into a suitably constructed furnace by means of a centrifugal fan where it became mixed with the gaseous fuel, and the products of combustion obtained, reduced to a temperature of 650° C. by excess of air, were "then employed in the form of a jet, to work a wheel like a water-wheel." This is a patent of considerable interest, for it is in every essential feature the same as the Armengaud-Lemale turbine of some sixty years later. What, however, is the most significant fact in the matter is the mention of the temperature, 650° C. (1,200° F.), at which the hot gases are to be admitted on to the turbine wheel; the mention of the "compressing fan" or rotary compressor is also significant. The author feels it necessary to put forward the above information with some reserve. The notice of Bresson's invention is taken from a most unreliable source, namely Bourne's "Steam, Air and Gas Engines," London, 1878. To find any particular specification among the files of the French patents in the English Patent Office is rendered extremely difficult owing to the absence of an adequate index.

In 1856 appeared Fernihough's "combined steam and smoke engine" (No. 13,281), an illustration of which is given in Fig. 90. The necessary pressure required in the furnace was to be brought about by the injection of water. This is rather in the nature of a retrogade step from the more advanced schemes of Barber and Bresson.

In 1846 Burdin proposed a similar scheme to that of Bresson nine years earlier, with the addition of the particular mention of a *multiple* rotary compressor in connection with a *multiple* turbine.

So far the types of gas turbines noticed have been the prototypes of the three main classes of gas turbines; the single-fluid, constant-pressure, the mixed-fluid constant-pressure, and the explosion gas turbine. In the same year in which Fernihough patented his "steam-and-smoke engine," A. V. Newton took out a patent (No. 1672) for "Obtaining

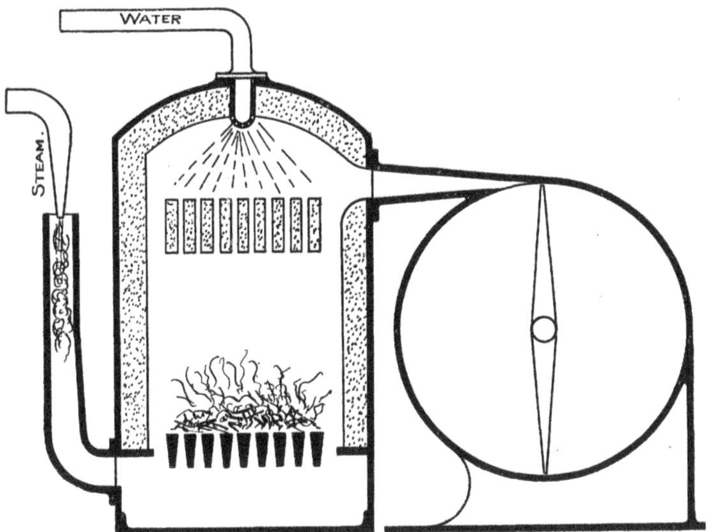

FIG. 90.—Fernihough's Steam and Smoke Engine.

Rotary Motion." Since the appearance of the Humphrey pump there have been numerous proposals in the air to work a "gas" turbine in the form of a water turbine, rotated by columns of water, which are kept in a state of oscillation by the explosion of gas upon the surface of the water. It is interesting to note that even this latest evolution of the gas turbine was suggested and patented fifty-seven years ago.

Newton concludes his specifications with the following words:—

"I wish it to be understood that. . . . I claim the method, substantially as described, of imparting rotary motion to a wheel, or wheels, by the pressure of steam *or other equivalent expanded gas* acting alternately on the opposite one of two columns of water or other liquid, con-

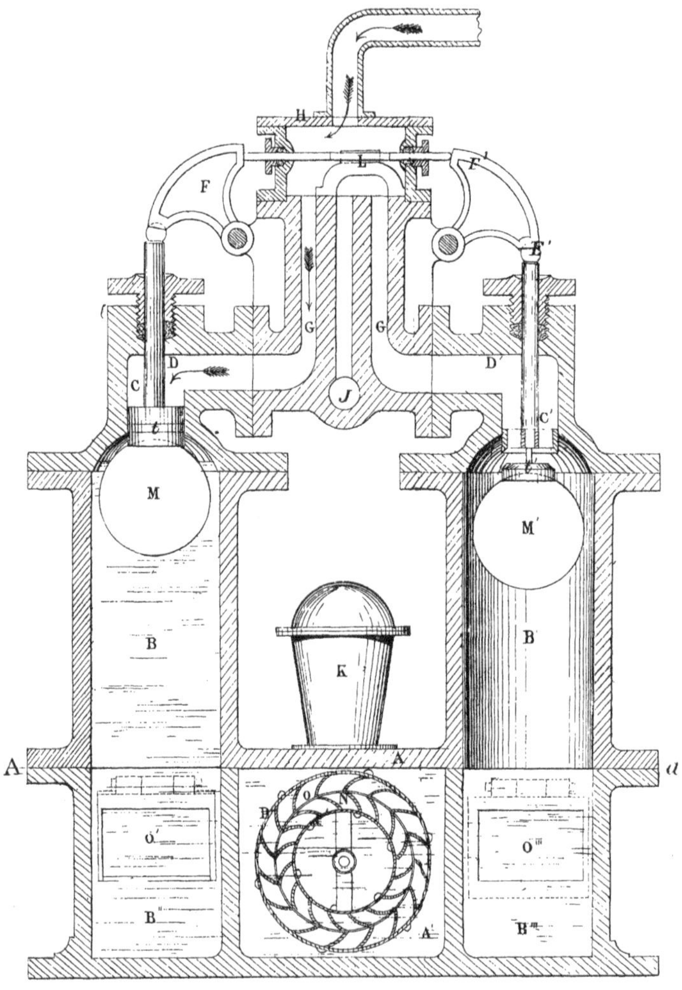

FIG. 91.—Newton's Gas and Water Turbine.

nected together, to cause the said water by such alternate action to pass through and impel the wheel or wheels, as set forth."

The machine is shown in Fig. 91. It explains itself.

The inventor primarily intended, of course, that his machine should work with steam ; indeed, he speaks of it in one place as an "hydro-steam engine." But the phrase "or other equivalent expanded gas" (the italics above are my own) presupposes the Humphrey pump gas turbine notion, though, it must be admitted, rather by implication than assertion ; rather by the desire to be comprehensive than from the wish to particularise.

Boulton, in 1864 and 1865, took out patents for the production of a working fluid under pressure, suitable for use in a gas turbine, and from this date various schemes—all more or less impracticable—for the use of the expansive power of gas upon turbine wheels, became fairly numerous. In 1872, however, a certain Dr. Stolze, of Charlottenburg, near Berlin, took out a patent for a "fire turbine." This machine was simply an externally-fired single-fluid, constant-pressure gas turbine. Such a type would be unworkable, even with the improved machinery at command to-day. It seems, however, that an actual machine of the kind was constructed.

The next patent of any importance is that of Charles Lemale in 1901. This brings us to the famous "pioneer" experiments of Armengaud and Lemale, which are discussed in the next chapter. Even the foregoing brief and incomplete survey of the history of the gas turbine shows one thing very clearly, that is, the practically negligible advance that has been made on the theoretical side of the question. The theory of the gas turbine was as fully grasped by Barber at the end of the eighteenth century, and by Bresson in the beginning of the nineteenth century, as it is by experts to-day. The success of the gas turbine as a heat engine rests *solely* upon practical limitations. Let a turbine and pump be so constructed as to give 75 per cent. thermodynamic efficiency ; let a wheel be made that will run at a permanent peripheral velocity of 900 ft./sec. and withstand a continued temperature of 500° C. ; let the machine and furnace be so efficiently water-jacketed as to reduce heat losses to 5 per cent., and we can produce a gas turbine to-day—as Barber could have done, under the same circumstances, in 1791—which will give an overall efficiency higher than any other known prime mover.

CHAPTER XI. THE PROGRESS IN EXPERIMENTAL WORK.

ONLY two—or, to be strictly accurate, only three—gas turbines have been actually built, made to operate successfully, and submitted to tests as to fuel consumption. These three experimental turbines are those of Armengaud and Lemale, of Karavodine, and of Holzwarth. The experiments of Armengaud and Lemale were made during the years 1903 to 1906 upon a 300 B.H.P. single-fluid, constant-pressure turbine, with steam injection ; those of Karavodine during 1908, upon an open-chamber explosion turbine giving 1·6 B.H.P. ; and those of Holzwarth during the years 1908 to 1911, upon a closed-chamber, constant-volume turbine of 1,000 nominal horse-power. The first and last of these three experiments are the only ones of real importance ; the Karavodine turbine, however will be briefly noticed in passing.

THE EXPERIMENTAL WORK OF ARMENGAUD AND LEMALE.

It is not a little difficult upon the evidence now to hand to decide whether this turbine should be classed as a single-fluid gas turbine in which a certain, but not a *maximum*, quantity of steam has been introduced for cooling purposes, or whether the turbine should be regarded as a steam-and-air turbine proper, in which all the available temperature reduction is effected by steam, the combustion mixture being of the richest possible composition. It would seem from the construction of the machine, and particularly that of the combustion chamber, that the former, rather than the latter, was the true nature of the turbine. The cycle was as follows :—

A Rateau rotary compressor, an illustration of which is shown in Fig. 79, compressed air from the atmospheric pressure to a pressure of 112 lbs. per square inch absolute, that is to say, a compression ratio of 7·6. The air was then introduced into a combustion chamber into which liquid fuel was injected. Part of the combustion chamber was water-jacketed, and steam raised from this jacket was admitted into the second half of the combustion chamber for the purpose of lowering the temperature of the working fluid upon the turbine blades. Adiabatic expansion was effected through an expanding nozzle—also water-jacketed—and the energy of the effluent gases absorbed upon a double wheel of the Curtis type. Armengaud mentions a system of regeneration by the raising of low-pressure steam, but it is not clear whether such a system was adopted in the experimental turbine actually built and tested in the shops of the Société des Turbomoteurs at Paris.

Before considering the construction of the machine, it may not be uninteresting to calculate the fuel consumption theoretically from the data of the practical limits within which the turbine in question operated.

The initial pressure, as before stated, was 112 lbs. abs., the expansion ratio being 7·6 ; the final temperature on the blades after expansion was 400° C. Disregarding the addition of the steam this makes the initial temperature of the working fluid—

$$\begin{aligned}\theta_1 &= \theta_2 \ . \ R^{\frac{\gamma-1}{\gamma}}\\ &= 673.\ 7{\cdot}6^{\cdot 275}\\ &= 1180\ {}^{\circ}\mathrm{A}.\end{aligned}$$

Therefore the work done—

$$\begin{aligned}W &= C_p(\theta_1 - \theta_2)\\ &= 127\ \mathrm{T.U.}\end{aligned}$$

The negative work done by the pump—

$$\begin{aligned}W &= 18{\cdot}7 \log_e 7{\cdot}6\\ &= 37{\cdot}2\ \mathrm{T.U.}\end{aligned}$$

Heat put in—

$$H = C_p(\theta_1 - \theta_0) = 218\ \mathrm{T.U.}$$

Therefore, thermal efficiency—

$$\eta = \frac{W - w}{H} = 41{\cdot}2 \text{ per cent.}$$

The Rateau pump used gave a thermodynamic efficiency of 65 per cent.

Let it be assumed that the turbine gave a thermodynamic

efficiency of 70 per cent. Then, overall efficiency, assuming no heat losses from furnace—

$$E = (89 - 57)/218 = 14{\cdot}6 \text{ per cent.}$$

From the diagram giving the efficiency curves for the single-fluid and mixed-fluid turbines (*vide* Fig. 34, p. 74) it may be seen that the addition to the air of the maximum quantity of steam for this temperature increases the efficiency something like 1·6 times. In the Armengaud turbine, however, the combustion chamber was not wholly water-jacketed, so that there must have been some considerable loss of heat from the combustion chamber ; nor was the steam which was added to the products of combustion used in the most efficient manner. We may, therefore, approximate matters by the following compromise of taking the heat losses from the combustion chamber to be 20 per cent., and the increase in efficiency due to the addition of steam to be 1·3. This gives an overall efficiency from fuel entry to turbine shaft of 15·1 per cent. ; or a fuel consumption of 0·937 lb. of oil per brake horse-power per hour. This is approximately the best which, under the conditions prevailing, could have been achieved ; the actual result was not so good.

The probable efficiency is herein worked out, because the actual results of tests are difficult of access. In an article in *Cassier's Magazine*, April, 1908, by M. Barbezat, it was stated that " these tests are not yet completed, and the results are not available for publication." The machine apparently never gave an adequate degree of success. This is very deeply to be regretted, as not only (in the writer's opinion at least) was the cycle adopted the right one, but much ingenuity and labour was expended on constructional details. A general view of the gas turbine itself, apart from the pump, is shown in the two photographs (Fig. 92, A and B). A section of the kind of combustion chamber used is shown in Fig. 93. The fuel is introduced through the pulveriser P ; the air through the entry A ; the ignition takes place at K. The further half of the combustion chamber is water-jacketed (to a limited extent) by the coil C, through which water is circulated, and the steam thus raised is introduced into the combustion chamber and mixes with the products of combustion. The description of the " atomiser " of Armengaud and Lemale is given on p. 155. The turbine

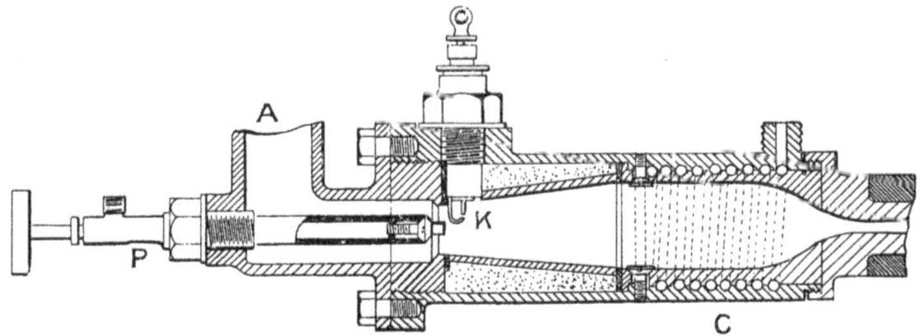

FIG. 93.—Armengaud and Lemale Combustion Chamber.

wheel which was used for these experiments was water-cooled, a system it may be noted in passing, which was not adopted in the Holzwarth experiments six years later. A section of the wheel is shown in Fig. 94. It will be seen that not only the disc upon which the blades were set, but the very blades themselves, were cooled by the internal circulation of water. The water ducts are shown in the drawing. X and Y are circular channels that run right round the rim; they were supplied with water by radial ducts-*z*; *x* and *y* are small ducts running up the middle of the turbine blades themselves.

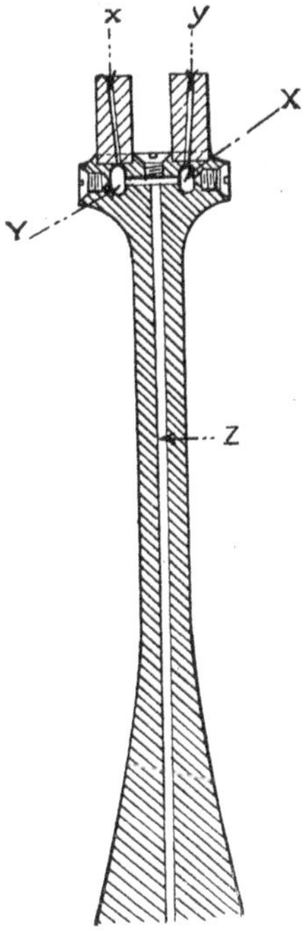

It was maintained that the difference in the specific gravity between the heated and the unheated water, together with the centrifugal force set up by the high rotational speed at which the wheel ran, secured an adequate circulation of the water. The combustion chamber appears to have been lined with carborundum to resist the high temperatures of combustion; an elastic packing composed of asbestos was employed at the back of this to allow the necessary give-and-take for the difference in the coefficients of expansion between the carborundum and the metal wall. This

FIG. 94.—Section through Water-cooled Wheel, Armengaud and Lemale Turbine.

does not seem to have been wholly a success, and it is reported that, after a short period of working, the carborundum lining was swept bodily through the nozzle into the turbine. The Rateau compressor was made to run at a speed of 4,000 r.p.m., and was coupled direct to the turbine. The compressor is said to have absorbed *one half* of the total power given out by the turbine. This shows where the inherent weakness in the turbine lay; the proportion of the negative work was too great. It would seem either that the initial temperature of the working fluid was not sufficiently high, or that the thermodynamic efficiencies of the turbine and pump were too low; or, what is more likely to have been the case, the proportion of water added to the working fluid was not enough, and even so the conditions under which it was added were not such as to conduce to the best results. That the peripheral speed of the turbine wheel was too low to admit of a proper absorption of power does not seem likely, considering the speed of revolution adopted.

It might here be pointed out that, had the turbine been worked as a sub-atmospheric turbine, expanding from 14·7 lbs. down to 1·93 lbs. abs., as might equally as well have been done —the pump being used as an exhauster instead of as a compressor—a much higher efficiency might have been attained; for not only the pump, but the turbine also, would have given greater efficiency owing to the decrease in frictional losses. Also several practical difficulties with regard to the combustion chamber would have been avoided, and the whole business of "burning under pressure" would have been eliminated.

The actual fuel consumption given by the Armengaud-Lemale turbine appears to be somewhat shrouded in gloom. It seems that M. Barbezat, who was associated with M. Rene Armengaud up to the untimely death of the latter, spoke of a fuel consumption of "1,200 to 1,300 grammes of petrol per horse-power hour." Taking the calorific value of the petrol to have been 10,000 T.U. per lb., this gives an overall efficiency of 9·3 per cent. This value is far below that calculated above of 15·1 per cent., though the usual codicil was added of "indication of the fuel consumption being very materially lowered."

It is a great pity that the cycle used by Armengaud and Lemale has not been revived by some experimenter under more favourable limiting conditions. Multiple compressors have to-day given efficiencies of over 70 per cent.; also the value of the temperature after expansion, that is to say, the temperature on the blades, has been raised. While Armengaud used a temperature only of 400° C. upon his blades, Holzwarth in his explosion turbine appears to have employed temperatures up to 940° C., though it is true that he advises nothing higher than 450° C.

THE EXPERIMENTAL WORK OF KARAVODINE.

In 1908, M. Karavodine constructed a turbine, in Paris, of the open-chamber type. It gave about 2 H.P., and operated

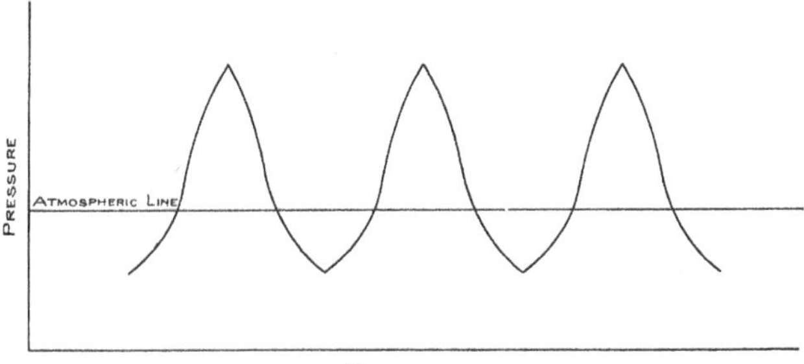

FIG. 95.—Time-Pressure Diagram, Karavodine Turbine.

with ease and regularity. The machine was extremely small. The wheel was of the de Laval type, and of some 6 inches in diameter, carried upon a flexible shaft. The turbine was fitted with four nozzles situated around its rim, to each of which was attached a separate combustion chamber. The form of the combustion chamber is shown in Fig. 69. X is a water jacket surrounding the body of the chamber; Y and Z are the fuel and air inlets respectively; D is a light valve held down by the spring A, and regulated as to lift by the screw B. A sparking plug is placed at C. The nozzle, N, which is shown broken, was 3 cm. in length and 16 mm. bore. The volume of the explosion chamber was 230 c.c.

Upon a charge being ignited in the chamber a sudden rise of pressure took place, and the gases rushed out through the nozzle, N, on to the turbine wheel. This outrush of the gases produced a momentary suction effect in the chamber. That is to say, the pressure therein fell below that of the atmosphere outside. This caused the suction valve, D, to open and to admit a fresh charge of combustion mixture. This, in its turn, was ignited by the sparking plug at C, which was timed to be in keeping with the periodicity of the explosions. Thus a continuous series of impulses was set up, and kept automatically continuous, without the mediation of any mechanically operated valve. The nature of the impulses produced is shown diagrammatically in Fig. 95. The co-ordinates represent pressure; the abscissæ, time. About thirty-eight explosions were made a second. The maximum pressure obtained was about 5 lbs. above the atmosphere; the maximum suction about 2 lbs. below. The wheel ran at a speed of 10,000 r.p.m., giving a peripheral speed of 258 feet per second. The machine in question was tested by M. Barbezat; it gave a fuel consumption of 5 lbs. of petrol per horse-power per hour, or an overall efficiency of 2·85 per cent.; a low figure, but it must be borne in mind that the machine was excessively small. A machine of this kind is so extraordinarily simple in construction and operation that, for small powers, there would seem to be no small opening for such, were an efficiency obtained approaching that of the small petrol motor.

THE EXPERIMENTAL WORK OF HOLZWARTH.

By far the most important experimental work that has yet been done in the elucidation of the gas turbine problem, apart from the historical estimate, was that executed in the last few years by Herr Holzwarth, of Mannheim. The final form that Herr Holzwarth's turbine took is shown in Fig. 96. The cycle here adopted was that of explosion in a closed chamber; that is to say, explosion behind a valve. Upon the explosion reaching its maximum pressure value the valve—which is held closed by a spring of the required strength—opens, and the products of combustion expand on to the turbine wheel. By an ingenious device the valve is held open while this efflux takes place, for otherwise the valve would close again by its spring directly the pressure fell to that value at which the opening of the valve first occurred.

The kinetic energy of the products of combustion are

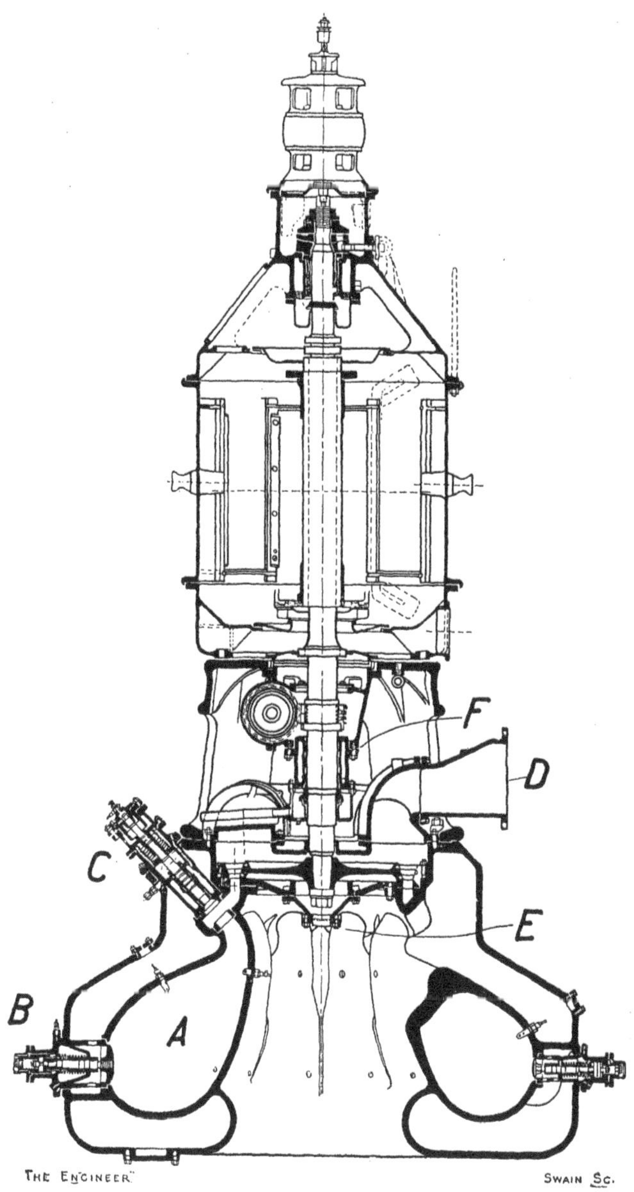

FIG. 96.—Holzwarth Gas Turbine (Vertical Section).

absorbed on a "double" wheel of the Curtis type. This wheel is of simple construction, mounted on a vertical shaft, and *not* water-cooled. The air and combustible was delivered into the chamber at an initial pressure higher than that of the atmosphere. The initial pressure in the combustion chamber varied from 1·3 to 1·9 atmospheres, the pressure upon explosion from 4·79 to 9·48 atmospheres. The initial compression was effected by means of a rotary pump of the Rateau type, driven by means of a steam turbine. The steam for this turbine was raised from the waste heat of the gases effluent from the turbine wheel. That there is ample waste heat available for such compression ratios as were used is shown in the theoretical analysis of this type of turbine in Chapter V. (*q.v.*). The experimental turbine shown in Fig. 95 was built for a nominal output of 1,000 H.P. The latest results to hand concerning the output and efficiency of this machine are those communicated to Mr. Dugald Clerk in connection with his paper upon the Gas Turbine, delivered before the British Association at Dundee, in September of last year. These were a maximum output of 450 B.H.P. and a maximum efficiency of 23 per cent. The absolute nature of this "efficiency" was not explained.

In the section through the turbine, Fig. 96, the explosion chamber is shown at A. A number of such chambers form a ring above which the turbine is centred and which forms a self-contained bed plate for the whole machine. The inlet valve is shown at B; there are two valves, one the inlet for the gas and the other for the air; C is the valve closing the combustion chamber when the explosion is to take place. After passing through the turbine wheel the exhaust gases are delivered through the orifice D to the steam regenerator. The dynamo is supported above the turbine, a foot bearing for the whole machine being provided at E. A sleeve surrounds the turbine shaft at F, on account of the fact that an exhauster is used to remove the waste gases and to assist in "scavenging," fresh air being drawn in to sweep out the burnt products that remain in the chamber after ignition and expansion has taken place. The speed of the turbine was 3,000 r.p.m. The diameter of the turbine wheel was slightly over 3 feet, giving a peripheral velocity of about 500 ft./sec.

The inlet valve is shown in section in Fig. 97. A is the valve shown at rest upon its seating. It is held firmly in place there by the spring B. C is a small piston fixed on to the valve spindle and butting against the spring B. This small piston works in the sleeve D, which is itself a piston working in the chamber F. Thus D and C form together a differential piston. The purpose of this contrivance is to facilitate the proper opening and closing of the valve A, not by any

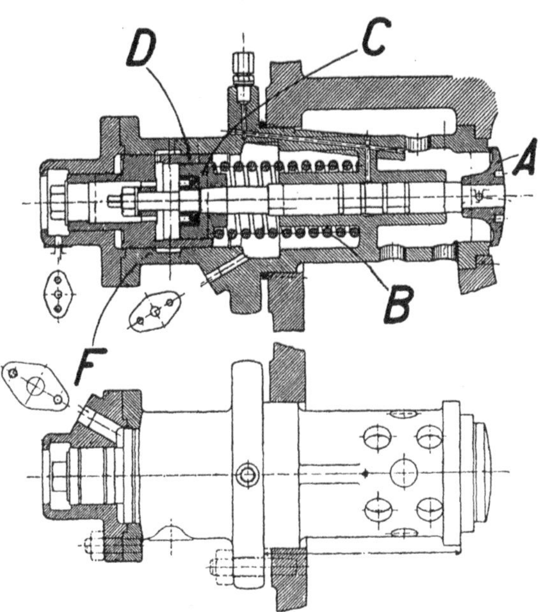

FIG. 97.—Inlet Valve, Holzwarth Turbine.

mechanical means but by a system of oil pressure-control, worked from a rotary distributor driven from the main turbine shaft. The use of the differential piston can be dispensed with, unless it is desired to control the valve from the outside. The oil pressure for opening the valve is admitted on top of the outer piston, a cushion of oil being held between the inner and the outer piston. During explosion, of course, the valve is held down by the pressure of gas in the combustion chamber as well as by the force of the spring.

The nozzle valve is shown in Fig. 98. A is the valve seating ; B the valve. The valve is shown in the Figure off its seating, in the position immediately after explosion. The opening of this valve is not controlled positively, either by oil control or mechanical means. The spring, S, holds the valve up against its seating until the ignition of the charge in the combustion chamber has taken place. The strength of the spring is so adjusted that when the force of the explosion reaches its maximum value—or, rather, just before that value is reached—the pressure of the explosion on the valve is sufficient to force it back against the spring. As soon as this

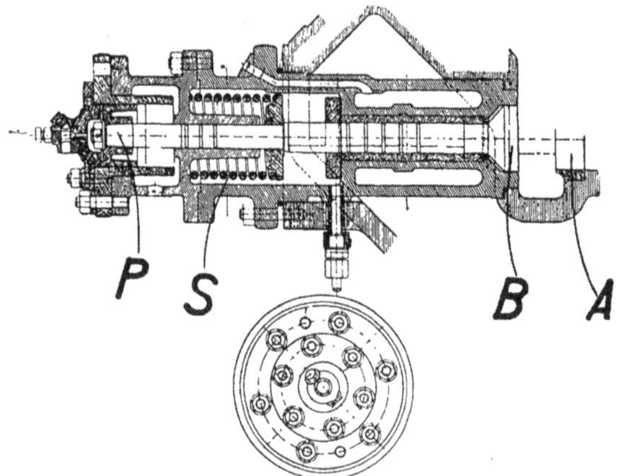

Fig. 98.—Nozzle Valve, Holzwarth Turbine.

has taken place, oil under pressure is admitted behind the piston, P, from the rotary distributor, and the valve forced fully open and held in that position until the gases in the chamber have fully expanded and the subsequent charge of scavenging air has swept through. The oil pressure is then released, and the valve closes by the air of the spring ; the combustion chamber is then ready for a fresh charge.

A section through the rotary oil distributor is shown in Fig. 99. It consists of a rotating drum, upon a vertical shaft, and is driven by a worm reduction gear from the main turbine shaft. There is a separate rotor for each of the three sets of valves ; namely, the nozzle valves, the gas-inlet valves, and the air-inlet valves. The oil is pumped under pressure into the space, A, above the rotor. The spaces X, Y and Z, feed oil to the three sets of valves as detailed above. The casing

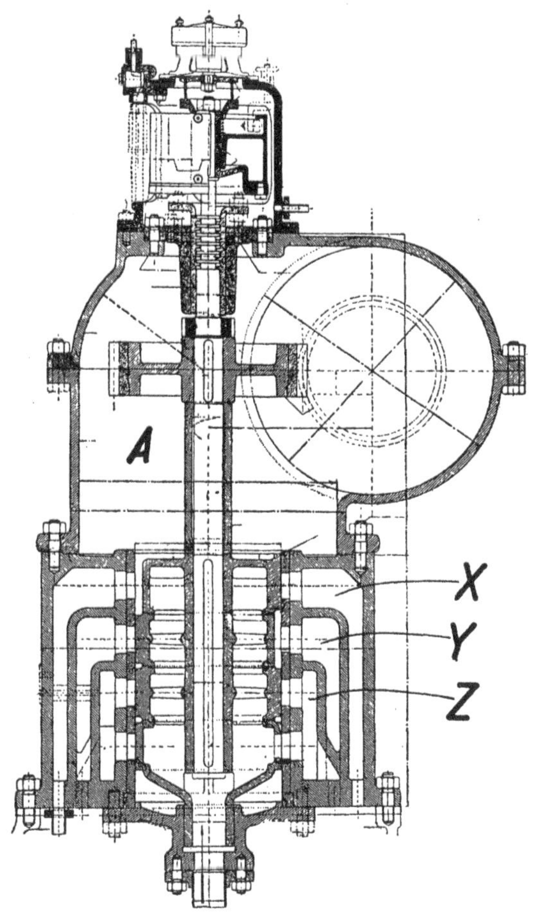

FIG. 99.—Rotary Oil Distributor, Holzwarth Turbine.

round the rotating drum is provided with axial ports, ten in number, ranged at equal intervals around the circumference. The rotor is itself provided with corresponding longitudinal ports. Oil is admitted under pressure to the interior of the rotor. It flows through the ports in the rotor, and thence through those in the surrounding casing, to the particular valve to be operated. When the particular port in the drum is again left open, the oil behind the piston in the valve chamber flows back into the oil sump, owing to the pressure of the spring closing the valve. The oil is then redrawn from the sump, compressed by the pump, and returned to the inside of the distributor at the required pressure.

The steam regenerator took the simple form of a long vertical fire-tube boiler, the hot gases entering at the top, traversing the regenerator to the bottom, and then ascending again to be delivered out at the top. A steam dome of a suitable nature was provided above the regenerator, from which the steam was taken to drive the blowing plant.

The blowing plant was of a simple nature, as the compression ratio used (maximum 1·9) was small. The plant consisted of a steam wheel, driven by the steam from the regenerator, a rotary compressor and an exhauster, or air circulator of the ordinary fan type, the pressure difference required to be set up by the exhauster being, of course, very slight. The blowing plant ran at a speed of 5,000 r.p.m. With rotors 2 feet in diameter this would give a peripheral speed for the blower of 520 ft./sec. This gives, according to the formula of Professor Rateau—

$$R = 1 + 6 \times 10^{-7} \, . \, v^2,$$

a compression ratio of 1·15, for a single element of the pump. Since we have the ratio—

$$\text{Number of Elements} = \frac{\log R'}{\log R}$$

where R is the compression ratio for one element and R′ is the total compression ratio, the number of elements required for 1·9 compressions would be from 4 to 5. This does not necessitate a large or cumbersome compressor. It may be borne in mind in this connection that the Rateau compressor used in the Armengaud-Lemale experiments was composed of twenty-five elements.

In the tests carried out on this turbine during the year 1911, the temperature of the working fluid after expansion, that is to say, the maximum temperature on the blades, was gradually, in successive tests, reduced to 400° C. This was the same value as used in the Armengaud-Lemale turbine. In Herr Holzwarth's article, in September, 1912, in the *Journal fur Gasbeleuchtung* (Munich), however, he states 450° C. as a practical working value for this temperature.

As regards the actual efficiency obtained by Herr Holzwarth, it would seem that there is some misunderstanding or uncertainty (perhaps through inadequate translation) about the matter. The following criticism in this connection is quoted from Mr. Dugald Clerk's paper on the Gas Turbine read before the British Association in September of last year.

> "Mr. Holzwarth appears to claim a possible conversion of 30 per cent. of the heat supplied to the turbine into mechanical energy at the turbine shaft, and I understand that he has made this claim for the present turbine in its imperfect state. There is some mistake about this, because even on the air standard cycle, taking the approximate compressions used, and assuming expansion to atmosphere, the efficiency is too low to permit in practice in a reciprocating engine with its smaller heat losses more than about 15 per cent. With the larger turbine losses the theory of the Holzwarth machine does not appear to me to permit more than a 10 per cent. heat conversion, and so far as I understand Mr. Holzwarth's results, his actual conversion is much less than this."

At the same time Herr Holzwarth telegraphed to Mr. Dugald Clerk that his best "efficiency" obtained up to date was "23 per cent." There is evidently some misunderstanding as to the reading of the word "efficiency."

In the Table of the results of the tests carried out on the 1910 turbine, and given at the end of Herr Holzwarth's book, the best "efficiency," denoted by the symbol "η_{total}," and not further defined, is given as 24·7 per cent. The initial compression in this case was 1·37; the initial temperature before ignition, θ_0, 288° A.; the temperature upon explosion, θ_1, 1,440° A.; the temperature after expansion, θ_2, 880° A. A value of 54·5 per cent. was given for "$\eta_{turb.}$," which, presumably, denotes the thermodynamic efficiency of the turbine. It is easy from these data to calculate the *overall efficiency* of the turbine.

We have the relation for this particular cycle that the work done, per lb. of working fluid—

$$W = C_v(\theta_1 - \theta_0) - C_p(\theta_2 - \theta_0).$$

Substituting the values of θ_1, θ_2, and θ_0, given above, we have—

$$W = 61{\cdot}2 \text{ T.U.}$$

Now the heat put into the system—

$$H = C_v(\theta_1 - \theta_0)$$
$$= 207 \text{ T.U.}$$

Therefore the thermal efficiency—

$$\eta = W/H = 29{\cdot}6 \text{ per cent.}$$

But the thermodynamic efficiency of the turbine, "$\eta_{turb.}$," equalled, apparently, 54·5 per cent. Therefore the overall efficiency of the plant (exclusive of gas producer and any heat losses under the temperature of 1,440° A.) must have been—

$$E = 29{\cdot}6 \text{ per cent.} \times 54{\cdot}5 \text{ per cent.} = 16{\cdot}2 \text{ per cent.}$$

How this can be made to coincide with the "η_{total}" of 24·7 per cent., as reported by Herr Holzwarth, it is difficult to see. The divergence can only be explained, either on the supposition that the generally adopted methods of calculating the thermal efficiency of heat engines do not apply in the case of the explosion turbine, or that by the phrase "η_{total}" is meant some other ratio than that of "work available on turbine shaft compared to the corresponding quantity of heat put into the system."

The value of 24·7 per cent. is given in Table I. of the Appendix to Herr. Holzwarth's book ; the Tables II. and III. give much lower results for "η_{total}" based on certain formulæ (Stodola) of calculation. Information that would be very acceptable would be a categoric statement of the actual *fuel consumption in lbs. of oil*, of stated calorific value, per brake horse-power per hour, given by the turbine, on test.

Apart from the question of efficiency there is one defect inherent in the Holzwarth turbine, and common to all gas-explosion turbines of the closed-chamber type, which imperils the practical success of the machine. This is the operation of the delivery valve from the combustion chamber.

It is extremely doubtful even whether explosion behind a valve which opens *outwards*, is ever likely to be continuously successful in practice, however ingenious—and Herr Holzwarth's ingenuity is worthy of the highest praise—the system of operating that valve may be. The valve must be opened by the pressure within the explosion chamber reaching a sufficiently high value. Any system of positively-opened emission valve is considered by Herr Holzwarth to be inadmissible in practice. The tension in the valve spring must be constant. It is not to be supposed that the pressure produced in the chamber by successive explosions will always be the same; any slight difference in the mixture admitted would cause a difference in the maximum pressure set up. This would seem to suggest the very probable occurrence of the opening of the valve being delayed unduly—a condition of affairs not likely to conduce to the highest efficiency. The strain on the valve after lengthy periods of working must be very great. The whole system of valve control, is, indeed, in the closed-chamber explosion turbine a very troublesome and complicated matter, even when operated upon so ingenious a plan as the oil control of Herr Holzwarth. Unless the explosion turbine is found to give a very much higher efficiency than the present adduced value, the inherent complications seem to be likely to hinder its successful competition with other less complicated, and probably more reliable, prime movers.

CHAPTER XII. THE FUTURE OF THE GAS TURBINE.

"To generalise," said Blake, "is to be an idiot: to particularise is the great distinction of merit." Yet there are cases and occasions in which to generalise is not only permissible, but the only possible course to adopt. It is not allowable to particularise about the future. Only the most general indications can be adventured as to the part that the gas turbine is likely to play among prime movers during the next few decades. Speculation is the most abortive of all the arts. But here, in conclusion, it may not be altogether out of place; and imagination, so strenuously—and, no doubt, so properly—excluded from scientific work, may be admitted for a while on sufferance.

Discountenancing any notion of using the Humphrey pump as a power unit to drive water turbines (a gas engine driving a centrifugal and the centrifugal driving a turbine would be more economical and less cumbersome—so why drag in the Humphrey pump ?), or any schemes of keeping bodies of water in continual motion by the explosive power of gas—for, so far, we are aware of no experimental data in regard to such devices—we are faced with the three possible types of gas turbines: the constant-pressure gas turbine, single-fluid or otherwise; the constant-volume gas turbine, with the closed combustion chamber, *i.e.*, the Holzwarth turbine; and lastly, the open-chamber or Karavodine turbine.

The constant-pressure gas turbines give the indication of the best fuel economy; the Holzwarth type a considerably lower economy; the Karavodine lowest of all. On the other hand, the mixed-fluid turbines would appear to be the most costly and cumbersome, the Holzwarth turbine less so; and the Karavodine, which can only be possible for very small powers, the simplest and cheapest of all.

From the most general standpoint, the advantages and

disadvantages of the various types of gas turbines may be summed up as follows :—

(A) SINGLE-FLUID, CONSTANT-PRESSURE GAS TURBINE.

I. Advantages.

(*a*) Economy : high, under conditions in which the limits of temperature, pressure, and thermodynamic efficiencies are also high.

(*b*) Absence of sulphuric acid difficulties owing to absence of steam in the working fluid.

(*c*) Continuous turning movement on shaft.

II. Disadvantages.

(*a*) Economy : low, if limiting conditions are not of the highest. Efficiency falls off very rapidly directly the temperature or thermodynamic efficiencies of turbine and pump begin to drop.

(*b*) Heat losses from combustion chamber tend to be heavy owing to impossibility of adequate water-jacketing.

(*c*) Difficulties with regard to an air-to-air regenerator, as compared to an air-to-steam regenerator.

(*d*) Difficulties of governing and control.

(*e*) Undue importance of rotary compressor owing to large proportion of " negative work."

(B) MIXED-FLUID STEAM-AND-GAS TURBINE.

I. Advantages.

(*a*) Economy : high. Falling off slightly with drop in limiting conditions. Therefore comparative unimportance of thermodynamic efficiency of pump, turbine, etc., compared to foregoing case.

(*b*) Minimum heat losses from combustion chamber and turbine owing to complete water-jacketing.

(*c*) Entire machine working under sub-atmospheric conditions ; extreme lightness of construction therefore possible. Maximum pressure in system not exceeding 14 lbs. per square inch.

(*d*) Ease in governing and control by use of additional high pressure wheel. (In this case the advantage of low pressure is annulled.)

(*e*) Reduced proportion of negative work, and consequent decreased importance of rotary compressor.

(*f*) Continuous turning movement on shaft.

II. Disadvantages.

(*a*) Probable presence of sulphuric acid in condenser owing to admission of steam into the system.

(*b*) Increased size of combustion chamber into what is practically a steam boiler.

(C) CONSTANT-VOLUME TURBINE, HOLZWARTH TYPE.

I. Advantages.

(*a*) Reduction of rotary compressor in size and importance as compared with the constant-pressure gas turbines.

(*b*) Ease of governing by the control of number of explosions per unit of time.

(*c*) Absence of any sulphuric acid difficulties (as compared to mixed-fluid turbines).

(*d*) Probable less heat losses from combustion chamber as compared to constant-pressure, single-fluid turbine.

II. Disadvantages.

(*a*) Low economy as compared to the two foregoing types.

(*b*) Greater heat losses from the combustion chamber as compared to the mixed-fluid turbine.

(*c*) Periodic impulses on wheel instead of continuous movement.

(*d*) The difficulty of getting a valve that has to be opened by explosion to work satisfactorily for any length of time.

(D) THE KARAVODINE TURBINE.

I. Advantages.

(*a*) Extreme simplicity; complete absence of positively-operated valves.

II. Disadvantages.

(*a*) Very low economy.

(*b*) Only suitable for very small powers.

This broadly represents a comparative survey of the three, or rather four, types of gas turbines that actual experiment has proved practically possible. It is clear that the mixed-fluid, constant-pressure turbine, has every advantage over the single-fluid, constant-pressure turbine, save in the respect of the sulphuric acid question. Constant-pressure turbines naturally commend themselves only for large powers. For small powers the machines are too cumbersome and too costly. It must be borne in mind that a rotary compressor of 500 H.P. is about the same size, and not considerably less costly than one of 2,000 H.P. For large power installations, where first cost is of small importance compared to economy in fuel consumed, the mixed-fluid, constant-pressure steam-and-gas turbine, would seem to be the most suitable.

For smaller powers, where a smaller and cheaper machine is desired, and where fuel economy is only of secondary consideration, the constant-volume turbine of the Holzwarth type is indicated. For very small powers, where extreme simplicity and lightness is desired above all else, and a large fuel consumption is of no moment, there would seem to be a field for the Karavodine open-chamber type of turbine.

It would seem, generally, that the mixed-fluid, constant-pressure gas turbine would find its most fitting field of operation in installations of 1,000 H.P. and upwards, that the Holzwarth turbine would be most suited to units from 1,000 H.P. down to 100 H.P., and that the Karavodine type would be eligible only for small powers, from 1 H.P. up to 20. There are certain special applications in which the gas turbine, were it once to become a success as a practical machine, might be employed. One of the most important of these is the utilisation of waste gases from blast furnaces. It is easily discernible how much better suited for this purpose is the gas turbine than the gas engine, if the former were as efficient as the latter. For this use the constant-pressure type of turbine chiefly commends itself. The advantage, in cases of large power installations, of using entirely rotary machinery and eliminating all reciprocating parts is patent. In a 3,000 or 4,000 H.P. unit of the mixed-fluid, constant-pressure turbine type there are no parts having reciprocating motion, no mechanically worked valves—in fact, no valves at all. With gas engines it is not often advisable to put more than 250 H.P. in one cylinder, and a gas engine plant of the same power might involve six or eight units with all the necessary and complicated valve mechanisms of the gas engine and all its inherent defects of strains in connecting rod and crank.

Another field for the gas turbine is the utilisation of crude oils, upon the spot where they are found, to generate electric power and distribute the same throughout the surrounding country. At the actual oil wells, where the fuel has to undergo no transit, the cost of burning oil fuel would be very much less than it is here in England ; and in any oil fields sufficiently near populated districts, where there would be the necessary market for the power, the setting up of central power installations might be very remunerative. In this connection the constant-pressure type of turbine in large power units would seem to be most suitable. The map (Fig. 100) is taken from Dr. Diesel's paper before the Institution of Mechanical Engineers, in March, 1912. It shows the distribution of the petroleum fields throughout the world in the year 1908. In places such as the oil fields of Pennsylvania and the shore of the Caspian sea, power stations of this sort might be set up of almost any size.

It might be possible to apply the constant-volume gas turbine to tractional purposes, though it is doubtful whether a small explosion turbine driving, as it would have to do, through much reduction gearing could equal in efficiency the ordinary petrol engine. The constant-pressure type would be likely to be banned from this and such similar applications, on account of the size of the rotary compressor.

The explosion turbine has been suggested for the aeroplane. If an efficient type of open-chamber gas turbine could be devised, such might be possible ; but the valve control in the closed chamber type would hardly seem to offer any advantage in simplicity over the petrol engine. Even in aerial work efficiency cannot be entirely disregarded, as the more efficient the engine used is, the less weight of petrol has the aviator to carry with him for any given distance of flight.

With regard to ship propulsion, there would not seem to be any particular likelihood of the gas turbine, in any form, competing with the Diesel oil engine. The oil turbine in its

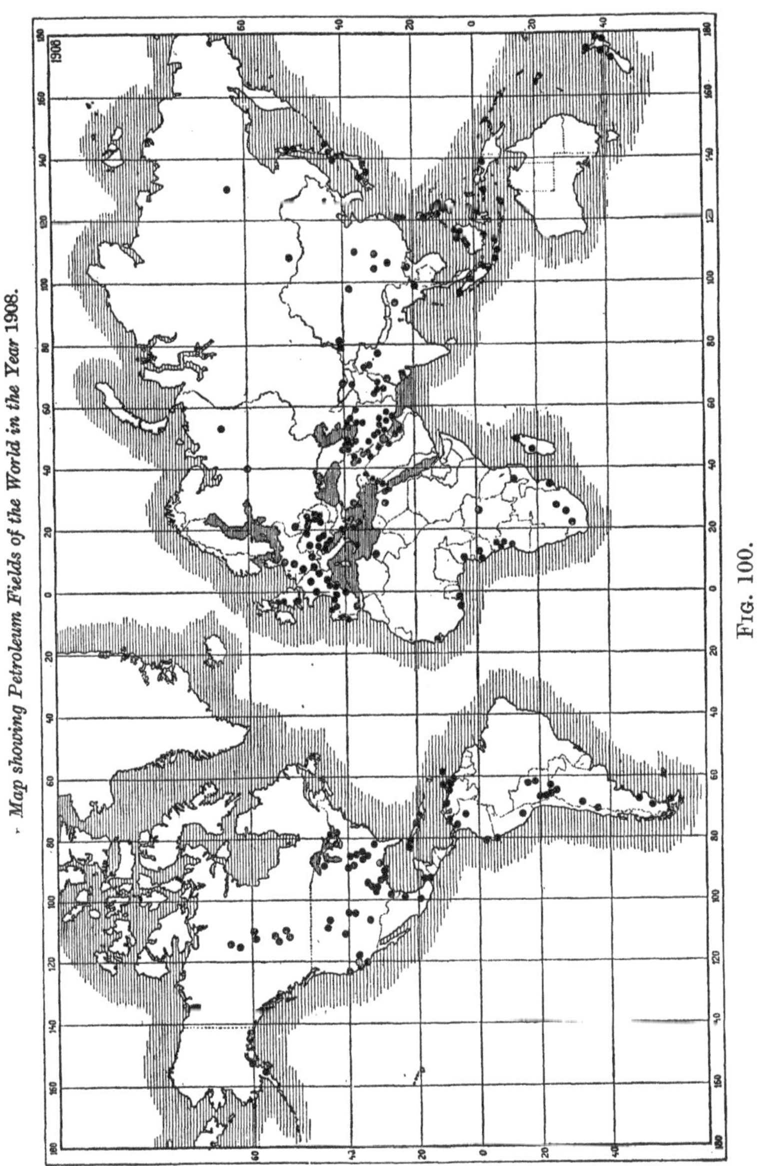

Map showing Petroleum Fields of the World in the Year 1908.

FIG. 100.

most efficient form, that is, the steam-and-oil turbine, does not promise more than a slight increase on the efficiencies already obtained with the Diesel engine, and with its rotary compressors the machine would be too bulky for marine work. In any case of ship propulsion where it was more reasonable to use oil instead of coal fuel the Diesel engine would be more suitable a machine to employ than any type of gas turbine. The proposals that have been made to use the Humphrey pump for ship propulsion seem sufficiently ridiculous to be dismissed with a reference as to the size and weight of this type of machine.

The chief field for the gas turbine in the future would undoubtedly seem to lie in the production of electric energy, whether oil, blast furnace, or producer gas be used. It must be borne in mind that the gas turbine is *essentially* a high-speed machine. As much as possible of the expansion ratio (in practice the whole of it) must take place in the first nozzle, which necessitates a high peripheral speed for high efficiencies. The speed cannot be reduced by any system of pressure step series, as is done in the Rateau or Parsons type of steam turbine. For this reason the gas turbine is only suited for the driving of dynamos or alternators, slower work requiring the interposition of gearing.

The foregoing survey of the future of the gas turbine is wholly speculative. Before concluding it may be not unedifying to place side by side the overall efficiencies that the various types of gas turbines ought to give and those which, in actual experimental work, have been recorded. These may be tabulated as on opposite page.

In the figures in the table it will be noticed that in 1905 Armengaud and Lemale only got half the efficiency on test that they should have obtained for the limits within which their machine operated. In 1911 Holzwarth approaches the theoretically calculated value very nearly. This has a special significance. In 1903—1906 the rotary compressor was not as efficient a machine as it is now. The combustion chamber used by Armengaud and Lemale was not so constructively perfect a piece of work as that of Herr Holzwarth in his explosion turbine. It is reasonable to suppose that had the experiments of the constant-pressure gas turbine had the benefit of the increased efficiency in accessory machinery and construction of to-day, results agreeing much more closely with the calculated values would have resulted. With regard to the Karavo-

	Calculated for same conditions as by experiment.	By experiment.	Date.
Single-fluid, constant-pressure turbine (Armengaud and Lemale).	15 to 20 per cent.	(300 H.P.) 9 per cent.	1903—1906
Single-fluid, constant-volume (Holzwarth)	12 to 16 per cent.	(1,000 H.P.) 10 to 15 (?)[1] per cent.	1908—1911
Mixed-fluid steam-and-air turbine.	20 to 35 per cent.	—	—
Open-chamber Explosion turbine (Karavodine).	—	(3 H.P.) 3 per cent.	1908

[1] *Vide* Chapter V., p. 118, and Chapter XI., p. 229.

dine experiment, the turbine was so small that anything approaching a reasonable economy was not to be expected. A repetition of the experiments of Armengaud and Lemale, and of Karavodine to-day, with improved machinery and on a more adequate scale, would yield no doubt very much better results. And there is still the mixed-fluid steam-and-gas turbine to be arraigned before the tribunal of actual experimental test.

One or two suggestions as to what would not be unwelcome in experimental data may not be out of place.

First, authoritative tests as to the efficiency of turbo-compressors, used as *exhausters* instead of compressors. Up to the present I am not aware of any such tests having been made. We know that a compressor of six elements, with a peripheral velocity of 850 ft./sec., will (according to Professor Rateau's formula) give a compression ratio of nearly ten. Will it do this as efficiently—or more efficiently, as theory indicates—below the atmosphere as above it ? A test of this nature would be very helpful and of the utmost importance to the success of the gas turbine.

Secondly, independent tests as to the temperature at which a bladed wheel can be run over considerable periods of time without undue erosion. By independent tests is meant tests as to temperature limits only, altogether apart from efficiency tests on any actual turbine.

Thirdly, an exhaustive series of tests as to the radiation and other heat losses from combustion chambers of various size, shape and material, with the combustion both at constant-pressure and at constant-volume, with special reference to the rate of increase of heat losses with rise in temperature.

Fourthly, investigations as to the behaviour of mixed fluids upon expansion.

Fifthly, inquiry as to whether the introduction of steam into the products of combustion of gas containing sulphur produces undue quantities of sulphuric acid, and, if so, means by which the evil may be remedied (*i.e.*, by introduction of base, etc.).

Sixthly, data as to the conductivity of heat, from air to air, through metal tubes.

Seventhly, experiments with explosions in open vessels and long tubes (*vide* Chapter V., p. 130).

An experimental investigation into any of the questions mentioned above would be invaluable to all interested in the success of the gas turbine as a prime mover.

It has been said more than once throughout this book that the gas turbine problem is entirely a practical problem. It would be only fitting and in order that I should end with a tribute to those who have been pioneers in experimental work; firstly, to the memory of Réné Armengaud, who, at a time when such was much more adventurous than it is to-day, built his 300 H.P. turbine in the face of many difficulties; and secondly, to Herr Holzwarth, who, as far as I know of, has been the only man since daring enough to adventure his opinion in as concrete a manner and on as large a scale.

The theorist inevitably feels that he is but repeating, or at the most interpreting, what others before him have said, or have tried to say.

Let it be his first duty to acknowledge his indebtedness to those who, sometimes with knowledge, often without, have had the courage to put their theories—and, incidentally, his—to the categoric test of experiment.

APPENDIX

LIST OF GAS TURBINE PATENTS, 1856—1913.

1856 .. 1,672 .. Newton.
1858 .. 2,271 .. Shaw and Cooper.
1860 .. 878 .. Henry.
1861 .. 1,633 .. Mennons.
1864 .. 1,636 .. Boulton.
1865 .. 1,915 .. Boulton.
1870 .. 1,859 .. Bourne.
1871 .. 2,326 .. Anderson.
2,587 .. Plessner.
1872 .. 1,155 .. Goransson.
1873 .. 2,246 .. Siemens.
1874 .. 486 .. Ford.
1875 .. 1,848 .. Clerk.
4,324 .. Preiswerk.
1876 .. 2,236 .. Martin.
2,386 .. Cook.
4,745 .. Cook.
1877 .. 4,571 .. Fallart.
1879 .. 1,161 .. Graddon.
1881 .. 2,280 .. Ford.
3,561 .. Kerkhove and Suyers.
5,237 .. Newton.
1883 .. 911 .. Capell.
1,552 .. Manghan and Waddy.
2,354 .. Trossin.
5,297 .. Wirth.
1884 .. 5,302 .. Johns and Johns.
9,394 .. Paddock.
11,361 .. Justice.
1885 .. 13,309 .. Dinsmore.
1888 .. 14,630 .. Susini.
1889 .. 4,302 .. Phillips.
5,559 .. Bergner.
1891 .. 741 .. Adams.
14,134 .. Watkinson.
1892 .. 13,939 .. Sayer.
1893 .. 6,204 .. Sayer.
1894 .. 7,485 .. Merryweather and Jakeman.
1894 .. 11,526 .. Redfern.
1895 .. 2,565 .. Ferranti.
9,922 .. Berrenberg and Krauss.
11,955 .. Atkinson.
23,740 .. Thomson and Webb.
1896 .. 180 .. Bollman and Kohnberger
1,404 .. Maxim.
2,436 .. Maxim.
6,073 .. Cook.
15,752 .. Gowlland.
16,630 .. Beetz.
19,134 .. Boult.
1897 .. 2,123 .. Martindale.
2,595 .. Ringlemann.
17,842 .. Marconnet.
19,673 .. Hayot.
28,821 .. Thompson.
29,508 .. Huber.
30,672 .. de Tangry.
1898 .. 7,398 .. Stolze.
19,392 .. Bäckerström
21,079 .. Vandel.
24,847 .. Coard and Charpentier.
26,767 .. Thrupp.
26,802 .. Edge.
1899 .. 11,433 .. Haddam.
11,544 .. Tygard and Tygard.
11,557 .. Weichelt.
17,721 .. Nivert.
1900 .. 4,938 .. Baillie.
5,970 .. Steffen.
8,052 .. Softley and Clements.
13,512 .. Steffen.
16,603 .. Windhausen.
23,479 .. Coleman.

Year	No.	Name
1901	1,071	van Beresteyn.
	2,858	Morin.
	6,469	Probst.
	12,696	Heinzeling.
	17,520	Edge.
	19,123	Rowbotham and Rowbotham.
	19,378	Pollak.
	19,920	Fullagar.
	22,131	Humphrey.
1902	5,711	Jacob.
	6,258	Enderly and Enderly.
	10,601	de Kierzkowski-Stewart.
	14,946	Porter.
	17,605	Loumiet.
	18,032	Hawkins.
	22,992	Cumming.
	24,781	Ferranti.
	25,179	Huskisson.
1903	4,758	Lemale.
	6,041	Kaudsen.
	6,287	Stippe.
	6,422	Lemale.
	7,685	Ferranti.
	13,199	Ferranti.
	14,308	Boult.
	14,936	Cumming.
	16,584	Smal.
	16,998	Godoy.
	17,560	Lees.
	18,459	Ashworth.
	21,886	Mars.
	23,872	Wilson.
	24,064	Drummond.
1904	1,092	Palairet and Barker.
	1,409	Ferranti.
	6,317	Fullagar.
	6,380	Smith and Lewis.
	6,570	Mackintosh.
	9,495	Ferranti.
	9,496	Ferranti.
	10,281	Thompson.
	12,239	Westinghouse and Leblanc.
	13,876	Semmler.
	14,385	Pichery.
	14,552	Wasley.
	14,781	Crossley and Atkinson.
1904	17,180	Stolze.
	17,296	Blieden and Davies.
	18,329	Hessler.
	18,404	Ker.
	18,435	Aktieselskabet Elbing Compressor Co.
	20,137	Young.
	21,320	Fullagar.
	22,001	Lemale.
	23,037	Eckersley.
	24,591	Fullagar.
	24,807	Knäpper.
	25,199	Windhausen.
	25,199A	Windhausen.
	26,178	Vogt.
	28,627	Thompson.
	28,833	Beck.
	29,495	Ferranti.
1905	3,607	Vogt.
	3,985	Lentz.
	3,985A	Lentz.
	6,682	Warick Machinery Co. (G. E. C.).
	7,847	Steffens and Rothe.
	8,831	Stodola.
	8,881	Stimner.
	10,204	Zoelly.
	12,769	Semmler.
	13,185	Lemale.
	13,352	Gasmotoren-Fabrik Dentz.
	13,473	Stodola.
	14,389	Renshaw.
	14,405	Gasmotoren Fabrik Dentz.
	14,486	Gasmotoren Fabrik Dentz.
	14,461	Greenwood and Anderson.
	14,461B	Greenwood and Anderson.
	14,461C	Greenwood and Anderson.
	17,941	Marcille.
	18,665	Smal.
	19,632	Semmler.
	19,912	Ferranti.
	24,320	Smith.
	24,656	Marquis de Chasseloup-Lanbat.
	25,049	Retz.

1905 .. 25,582 .. Fullagar and Bottomley.
26,699 .. Gelleri and Pernyss.
26,950 .. Wilson.
27,019 .. Villopique.
27,272 .. Gadda.
1906 .. 2,355 .. Hutchings.
2,818 .. Knauff.
3,570 .. Winand.
3,962 .. Hutchings.
5,320 .. Meyersberg.
7,603 .. Rollin.
8,273 .. Knauff.
8,886 .. Ferranti.
10,204 .. Karavodine.
10,376 .. Rateau.
11,874 .. Maschinenbau-Anstalt, Humboldt and Schmick.
12,055 .. Cruyt.
12,157 .. Ulmer.
12,735 .. Fullagar.
14,166 .. Ransford.
14,692 .. Ferranti.
14,841 .. Wardale.
15,776 .. Meyersberg.
15,786 .. Meyersberg.
15,787 .. Meyersberg.
16,267 .. O'Toole.
20,546 .. Holzwarth and Junghaus.
21,063 .. Seyboth and Baumann.
23,074 .. Wedekind.
23,123 .. Ghelli and Cazzani.
24,980 .. Büchi.
25,338 .. Bower.
26,006 .. Ashton.
26,725 .. Wedekind.
26,750 .. Brown, Barlow, Dvorkovitz and Olai.
29,022 .. Ayton.
29,589 .. Lorrain.
1907 .. 1,660 .. Rollin.
1,831 .. Eder.
5,872 .. Hutchings.
5,872A .. Hutchings.
6,977 .. Fennel.
7,086 .. Steffens.
7,912 .. Armengaud.
1907 .. 8,578 .. Holzwarth and Junghaus.
9,919 .. Smal.
13,852 .. Soc. des Brevets Étrangèrs.
14,031 .. Barclay.
15,127 .. Rollin.
24,085 .. Davey and Davey.
24,403 .. Soc. des Turbo Moteurs à Combustion.
24,842 .. Brooke and Brooke
26,184 .. Soc. des Turbo-Moteurs à Combustion.
26,215 .. Hutchings.
26,580 .. Davey.
27,219 .. Dodd.
27,287 .. Pécheur.
27,724 .. Esnault-Pelterie.
1908 .. 2,209 .. Esnault-Pelterie.
3,484 .. Kause.
8,321 .. Le Compagnon.
9,591 .. Sturm.
11,789 .. Piestrak.
11,868 .. Scheurmann.
11,915 .. Crabb.
12,829 .. Roberts and Roberts.
19,124 .. Hutchings.
23,493 .. Pazzeller.
24,309 .. Morgan and Kemp.
26,070 .. Lloyd.
27,622 .. Harrison and Coney.
1909 .. 530 .. Ferranti.
531 .. Ferranti.
607 .. Bour.
15,565 .. Holzwarth and Junghaus.
16,766 .. Dodd.
17,734 .. Meyersberg.
18,462 .. Illy.
20,668 .. Storey and Kitchen.
27,090 .. Lareinty-Tholozan.
28,577 .. Fessenden.
29,854 .. Lawson.
30,598 .. Holzwarth and Junghaus.
1910 .. 1,855 .. Gyrostatic Turb., Ltd., and Holroyd.

1910 .. 1,857 .. Gyrostatic Turb., Ltd., and Holroyd.
1,861 .. Gyrostatic Turb., Ltd., and Holroyd.
4,090 .. Lake.
6,185 .. Haubner.
7,375 .. Jourdain.
8,186 .. Praetorius.
10,946 .. Raclot and Enderlin.
14,311 .. Corthésy, Corthésy and Dickson.
15,576 .. Meden.
16,754 .. Little and Walker.
17,728 .. Freytag and Baumann.
19,175 .. Patschbee.
20,205 .. Rother.
20,994 .. Lethwaite.
21,094 .. Portass.
25,353 .. Dyer, Ritchie, Ritchie and Backstrom.
27,087 .. Wittgenstein.
27,292 .. Bour.
27,848 .. Lieblich.
1911 .. 900 .. Holzwarth and Junghaus.
4,518 .. Heilmann.
4,793 .. Philbrow.
1911 .. 4,998 .. Whiteside.
5,361 .. Raclot and Enderlin.
6,265 .. Hebden.
8,444 .. Corthésy.
10,462 .. Shouls and Allcock.
11,289 .. Blake.
11,008 .. Imhoff.
12,218 .. Castelli.
16,613 .. Baxter.
17,338 .. Sayer.
18,055 .. Kutschinski.
18,368 .. Sauer.
18,821 .. Coster.
19,925 .. Whiteside.
20,471 .. Hildebrand.
20,142 .. Brown.
21,268 .. Holzwarth and Junghaus.
1912 .. 4,350 .. Wedekind.
4,351 .. Wedekind.
6,968 .. Kothe and Gumpel.
8,131 .. Horch et Cie.
8,814 .. Josse.
7,733 .. Imrig and Imrig.
18,415 .. Pope.
1913 .. 503 .. Lemale.
1,390 .. Graemireg.
11,723 .. Kothe.

INDEX

BRADBURY, AGNEW, & CO. LD., PRINTERS, LONDON AND TONBRIDGE.

www.ingramcontent.com/pod-product-compliance
Lightning Source LLC
LaVergne TN
LVHW091643100826
845152LV00006B/143/J

* 9 7 8 1 9 2 9 1 4 8 2 0 2 *